The British Lighthouse Trail

Published by
Whittles Publishing Ltd.,
Dunbeath,
Caithness, KW6 6EG,
Scotland, UK

www.whittlespublishing.com

ISBN 978-184995-440-2

Printed and bound in Great Britain
by Severn, Gloucester

Cover image: Dubh Artach
Back cover images:
Noirmont Point, Channel Islands
Herd Groyne, Tyne and Wear
Maidens West Tower, Northern Ireland
Port Glasgow Beacon, Scotland
Opposite: Sule Skerry

The British Lighthouse Trail

A Regional Guide

Sarah Kerr

Whittles publishing

CONTENTS

INTRODUCTION

Back in 2012, I undertook a month-long tour of lighthouses on mainland England, Scotland and Wales. After months of planning I set off with, what I thought was a comprehensive list of lighthouses, but I found a number of others on my journey, particularly in Scotland.

After my tour, I discovered even more I had missed which encouraged me to continue researching. This research expanded to incorporate island lighthouses as well as those in Northern Ireland, the Isle of Man and the Channel Islands.

Whilst there are some printed publications and online lists available most are either fairly geographically selective, include only the larger structures, or feature almost every lit aid to navigation.

I felt that those wanting, as I did, to visit all of the British lighthouses, or those interested in a specific area, really needed a handy resource to help plan, and take with them, on their adventures.

I hope that this book will prove useful to individuals who purposely take trips to visit lighthouses and wish to have a guide to help along the way, and equally to those who are interested in seeing lighthouses in the areas they happen to be visiting.

Defining a 'visit' to a lighthouse is down to personal choice or capability. To one person it may be simply seeing the lighthouse from any distance or taking a picture of it, while another person may consider touching the building or getting inside to be a true visit. Membership of the Association of Lighthouse Keepers (www.alk.org.uk) is valuable for anyone who enjoys visiting lighthouses. They frequently organise special events and at times can arrange group access inside some lighthouses that are not usually open to the public.

This book is a regional guide to lighthouses throughout the mainland and islands of England, Scotland, Wales, Northern Ireland, the Isle of Man and the Channel Islands. I hope it helps you on your lighthouse trail as you travel to many of our nation's iconic, unique and awe-inspiring maritime aids to navigation.

ACKNOWLEDGEMENTS

I am grateful to those who have contributed to this book as well as the journeys it has taken me on.

This book would not exist if it were not for my husband's encouragement, support, and ability to 'make it happen', while never once suggesting we take a break from lighthouses!

I owe a great deal to John Best who has been incredibly generous with his time, knowledge, pictures and advice, as well as responding to my frequent questions and frustrations.

Thanks to Brian Johnson for being the oracle of all Shetland lighthouse knowledge.

I am thankful to Ian Duff for his final checks and meticulous attention to detail.

To those who have offered advice and information, contributed pictures and responded to my enquiries, I am enormously grateful.

Huge thanks to the team at Whittles Publishing for taking on the concept and seeing it through to completion and beyond. It has been a real pleasure to work alongside them.

DISCLAIMER

Some important points for readers to note before using this book:

1. This list is, by no means, the be all and end all. It is based, to the best of my knowledge, on information available and the situation in early 2019. I am open to comments and suggestions, particularly if you feel a lighthouse has been missed or should be excluded. Feedback received will be considered for any future edition. Feedback should be submitted at www.uklighthousetour.com/contact-me.
2. While this book contains directions and grid references to help you in visiting the lighthouses, reasonable care, attention and consideration should be paid to accessing these lighthouses, a number of which are on remote or tidal islands or private property. Where possible I have tried to include this information. Many offshore lighthouses no longer have well-maintained landing points or paths and any attempted landings are undertaken entirely at your own risk.
3. The pictures featured in this book are my own unless otherwise credited and should not be re-printed, re-used or copied without prior permission. Requests to use any images should be submitted via the website link above.
4. This book should not be used to assist maritime navigation. For this purpose, the latest Admiralty List of Lights should be obtained.

WHAT IS A LIGHTHOUSE?

To some the question above will seem almost absurd. To others who have perhaps considered embarking on a venture to visit all British lighthouses, it will be a reasonable question. The following quotation from D. Alan Stevenson's publication *The World's Lighthouses from Ancient Times to 1820* is a more than adequate explanation of the problems associated with trying to define a lighthouse:

> "...the term *lighthouse* cannot be defined exactly and the apparently simple and obvious definition as 'a structure bearing a light to assist navigation' is too wide: the varieties of navigation lights extend from the simple hand-lantern which guides a boat to its haven or indicates a particular state of the tide, to the elaborate sea-tower of the present day giving powerful light-flashes as well as other signals. Some towers with outlines that are supposed to denote a lighthouse bear a light of little power or value for shipping while other structures without resemblance to the accepted lighthouse-form exhibit powerful lights of great importance. So one cannot draw a sharp line of demarcation between powers of lights or types of structures and must just accept the fact that to some extent the term *lighthouse* is used arbitrarily."

However, in the interests of preparing a list of lighthouses it was necessary to set specific criteria.

The definition used for the purposes of this book is:

> ***"A fixed structure that was built to exhibit a light for the purpose of aiding maritime navigation and allows access for at least one person inside any part of it."***

This definition pays no consideration to the present or historical range of the light.

Included are:

- Remains of any old lighthouses if they are still large enough for a person to fit inside.
- Structures built to serve as a lighthouse that were never lit.
- Lighthouses that have been relocated from their original location.

Excluded are:

- Daymarks that have never exhibited a light.
- Light vessels, or any other floating aid to navigation.
- Solid towers with no access inside any part.
- Structures from which the only enclosed section has been removed, or the access point has been replaced/blocked.
- A small light with no internal access, mounted on top of a pre-existing building that was not originally constructed to feature a light.
- Lights erected for the purpose of aiding aviation or other forms of navigation aside from maritime.

Picture credits

All pictures are my own with the exception of the following:

John Best: Muckle Skerry (page 9), Out Skerries (10), Muckle Roe (19), Ness of Hillswick (20), Foula (21), North Ronaldsay (23), North Ronaldsay Old (24), Start Point (25), Noup Head (25), Auskerry (26), Helliar Holm (27), Copinsay (28), Tor Ness (32), Rubha Cadail (37), Sgeir Ghlas (53), Ardtreck Point (59), Uiginish (60), Vaternish (61), Rona (62), Gairlochy (71), Ardmore Point (92), Sgat Mor (109), Caladh Beacon (110), Carraig nan Ron (110), Rubh'an Eun (114), Pladda (117), Tayport High (123), Tayport Low (124), Newhaven Harbour (132), Port Logan (139), Little Ross (140), Ramsey (156), Laxey (157), Derbyhaven Breakwater (158), Point Lynas (166), Skokholm (175), Watwick Point Rear (176), Great Castle Head Rear (176), Whiteford Point (178), Mumbles (179), Monkstone (181), Charston Rock (183), Redcliffe (183), Whitehaven (186), Whitehaven Old New Quay (187), Foulney Island (189), Near Naze (190), Plover Scar (191), Ellesmere Port (192), Hilbre Island (195), Longstone (200), Coquet (201) North Shields High Old (202), North Shields Fish Quay High (203), North Shields Fish Quay Low (203), North Shields Old Low (205), Whitby East Pier (211), Whitby East Pier Extension (212), Whitby West Pier (212), Whitby West Pier Extension (212), Flamborough Head (214), Spurn Point (215), Thorngumbald (216), Whitgift (217), River Ouse Apex (217), Killingholme North Low Old and Killingholme High (218), Killingholme South Low (218), Slimeroad Front (229), Slimeroad Rear (230), Inward Rocks Front (230), Sharpness Docks (230), Berkeley Pill Front (231), Berkeley Pill Rear (231), Fishinghouse Front (232), Avonmouth North Pier Head (233), Avonmouth South Pier Head (233), Watchet (236), Instow Rear (239), Gravesend Royal Terrace Pier (245), Folkestone Harbour (250), Dungeness old (251), Portreath (254), Godrevy (254), Hayle Rear Old (255), Longships (256), Wolf Rock (256), Lizard (258), Fowey Whitehouse Point (260), Polperro (261), Hooe Lake Point (261), Eddystone (267), Plymouth Breakwater (268), Mount Wise Pier (268), Kingswear (270), Portland North Eastern Breakwater (273), Hurst Point (275, 276), Nab Tower Old (277), Nab Tower (279), Royal Sovereign (283), Hastings West Hill (283), Alderney (285), Casquets (286), Platte Fougere (287), Saint Sampson Harbour (287), St Peter Port Rear Belvedere (289), St Martin's Point (289), Les Hanois (290), Point Robert (291)

Andreas Herfurt: Sule Skerry (page iii and 33)

Brian Johnson: Lunna Holm (page 9), Ve Skerries (18)

Richard Lafferty, Aquaholics: Rathlin West (page 146)

Douglas Law: Fair Isle North (page 21), Weaver's Point (55), Eilean Trodday (61)

Joy Tubby: Happisburgh (page 222)

USING THIS BOOK

Traditional lights

Within the listing the names of many lights feature a star alongside. These are the structures that resemble a traditional lighthouse in their current state, with a typical lantern room and, in most cases, a tower. All other lights are listed without a star, including those which are now in ruins.

Lighthouse positioning

Each lighthouse listed falls within one of 5 categories. These categories are marked with abbreviations, which refer to the following:

- *m* – mainland: located on the mainland of England, Scotland, Wales or Northern Ireland. It should be noted that islands, such as Skye and Anglesey, which are now connected to the mainland by a bridge are also listed as mainland in the interests of accessibility from mainland UK.
- *i* – island: located on an island off of mainland England, Scotland, Wales or Northern Ireland. This category also includes lighthouses on man-made breakwaters that are not connected to the mainland, or on forts or platforms at sea.
- *t* – tidal: accessible at low tides from mainland England, Scotland, Wales or Northern Ireland.
- *i-t* – island tidal: accessible at low tide from an island.
- *r* – rock: located on rocks which are routinely submerged by the sea.

Location

A description of the location of the lighthouse. If the lighthouse has been relocated then details are included here and, where possible, I have included their original location. If a light has been relocated since it was discontinued its current location reflects where it appears in the list.

Description

A description of the lighthouse, including its material and colour as well as its height.

Established

When the current lighthouse was established and not when the location was first marked by a lighthouse.

Designed by

Where lighthouses have a particularly notable designer this information is included.

Admiralty number

Every active lighthouse in this book has a unique identity number assigned by the Admiralty. While this information may not help to identify lighthouses from the land, regular boat operators will be familiar with the Admiralty List of Lights, which contains all current active lights. Most active Northern Lighthouse Board lighthouses feature a name plate, often close to the main entrance door, which also contains the Admiralty number for that particular lighthouse.

Light characteristic

Lighthouses are identifiable by the characteristic of their light, for example how many times they flash and how often, whether the light is white, red or green, etc. Throughout the book I have used common abbreviations for the characteristics. The following list describes the abbreviations used:

- Fl: Flashing – light is mainly off with flashes of light
- LFl: Long flashing
- Q: Quick flash
- F: Fixed light
- Oc: Occulting – light is mainly on with shorter periods of darkness
- Iso: Isophase – light is on and off for equal lengths of time
- Al: Alternating – light alternates between two colours
- Mo(D): a series of light flahes representing the letter 'D' in Morse Code, with a 1-second flash signalling a dot and a 3-second flash a dash

If a number appears in brackets after the abbreviation this denotes the number of flashes (for flashing characters) or periods of darkness (for occulting characters) during the given period of time. If no number is given then there is only one flash or period of darkness.

The characteristic also includes the colour, or colours, of the light. These colours are abbreviated as follows: W – white; R – red; G – green; Bu – blue and Y – yellow.

When colours appear as "WRG" it is often a light made up of sectors.

5s/8s/10s: refers to the amount of time between flashes or darkness, so these examples would refer to 5 seconds, 8 seconds or 10 seconds.

Some examples:

- Fl WRG 6s: Light flashes white, red and green every 6 seconds
- Fl(2)W 10s: Light flashes twice, white light, every 10 seconds
- F R: Fixed red light
- Oc W 5s: White light goes off every 5 seconds
- Iso G 4s: Green light is on for four seconds and off for 4 seconds
- Al WR 8s: Light alternates between white and red every 8 seconds

Grid references

Grid references can either be used alongside Ordnance Survey maps, Global

Positioning System (GPS) devices, or a number of websites or mobile device applications to help you find the lighthouses listed.

For England, Scotland, Wales, Northern Ireland and the Isle of Man six-figure British National Grid references are used. While these grid references aren't as accurate as ten-figure grid references would be, they should take you close enough to the lighthouse to be able to see it. As the Channel Islands do not feature on the National Grid, latitude longitude coordinates in degrees and minutes have been used.

Access inside tower

Indicates whether or not the tower is open to the general public. A number of the towers are open only on Open Doors days or similar annual events. If this is the case the listing states that it is open only on special occasions.

Access to site

Advice on where to best access the lighthouse from. If a lighthouse is particularly hard to reach key viewpoints from which it can be seen have been given.

It should be noted that for island groups that feature outlying islands of their own - such as Shetland, Orkney, the Western Isles, the Isle of Man and the Channel Islands – the access advice given is from the main island of that particular group.

All distances included in the book are one way and should be doubled to determine the total distance if returning by the same route.

If any lighthouses are close together, details of how a visit to both in the same trip can be achieved have been included.

Details of lighthouse accommodation is not included in this book. *Lighthouse Accommodation: Britain and Worldwide* by Joy Adcock is a valuable resource for this purpose and can be ordered at http://lighthouseaccommodation.co.uk.

This information should be used as a guide only and, as land ownership and access can change, you should remain vigilant to any signs or local advice.

Consideration should be given to local residents and/or workers when visiting lighthouses in built-up or harbour areas. Where possible advice is given on nearby parking. Where there is no mention of parking you should use your own discretion. If you are unsure about parking in a particular area it may be appropriate to check locally before leaving your vehicle.

A number of the lighthouses are on breakwaters or piers. Care should be taken when accessing these structures and walking to them should be avoided in rough sea conditions.

Due to the remote location of many lighthouses, the surrounding areas are often home to a greater number of birds during nesting season. Many of these birds can be aggressive when protecting their nests and care should be taken to avoid disturbing them where possible at these times. Livestock should also be given space and respect particularly when with their young.

Additional notes

Notes have been supplied for certain entries. These relate to the following: the current whereabouts of any tower formerly in that location where applicable or known; where the former lens or lantern from the lighthouse can be seen if applicable or known; if another notable aid to navigation or associated building that doesn't meet the criteria outlined in this book can be seen nearby; or additional information that may be of interest.

Regions

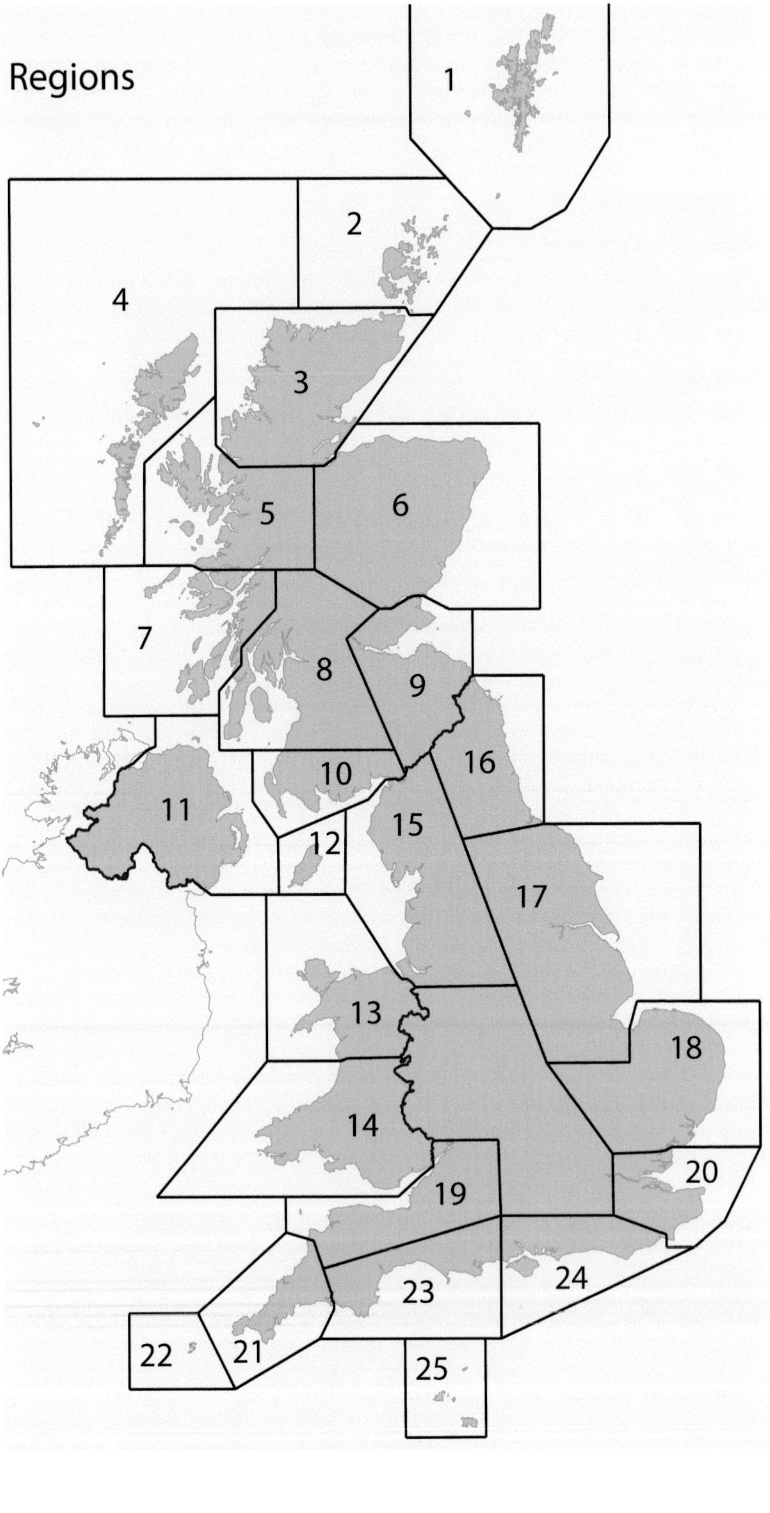

INTRODUCTION TO THE TRAIL

This book features 612 lighthouses.

294 of these lighthouses are in Scotland, 205 in England, 41 in Wales, 27 in Northern Ireland, 23 on the Isle of Man, and 22 in the Channel Islands.

For ease of use, the list has been divided into regions with each region home to the following number of lighthouses:

Region 1 - Shetland: 49 lighthouses

Region 2 - Orkney: 21 lighthouses

Region 3 - Northern Scotland: 20 lighthouses

Region 4 - Western Isles: 23 lighthouses

Region 5 - Inner Hebrides and West Highlands: 26 lighthouses

Region 6 - North East Scotland: 41 lighthouses

Region 7 - Mull to Islay: 34 lighthouses

Region 8 - Kintyre and Firth of Clyde: 36 lighthouses

Region 9 - South East Scotland: 34 lighthouses

Region 10 - South West Scotland: 10 lighthouses

Region 11 - Northern Ireland: 27 lighthouses

Region 12 - Isle of Man: 23 lighthouses

Region 13 - North Wales: 13 lighthouses

Region 14 - South Wales: 28 lighthouses

Region 15 - North West England: 28 lighthouses

Region 16 - Northumberland and Durham: 31 lighthouses

Region 17 - Yorkshire and Lincolnshire: 21 lighthouses

Region 18 - East Anglia: 17 lighthouses

Region 19 - Bristol Channel and North Devon: 27 lighthouses

Region 20 - Thames Estuary and Kent: 21 lighthouses

Region 21 - Cornwall: 20 lighthouses

Region 22 - Isles of Scilly: 4 lighthouses

Region 23 - South Devon and Dorset: 15 lighthouses

Region 24 - Hampshire, Isle of Wight and Sussex: 21 lighthouses

Region 25 - Channel Islands: 22 lighthouses

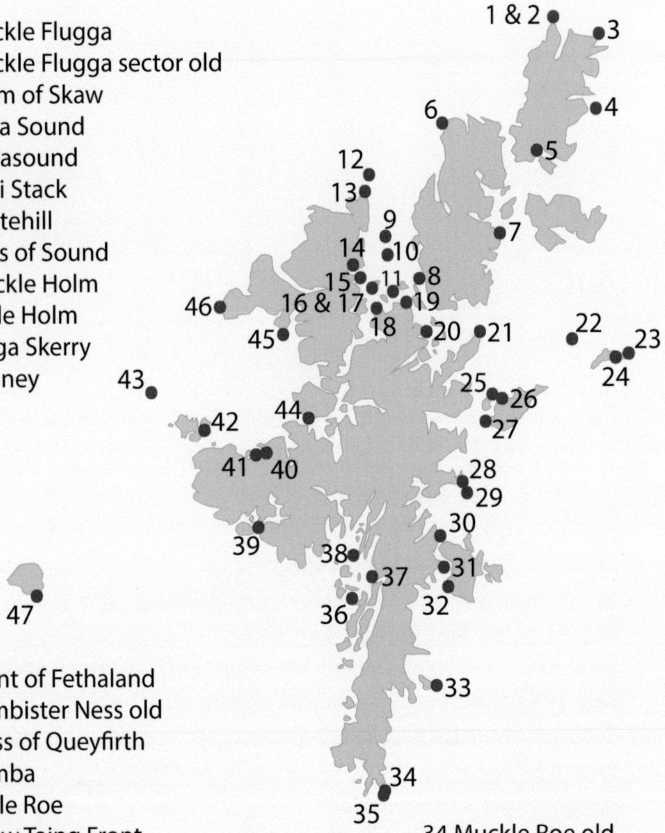

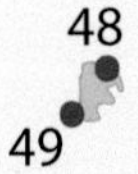

1 Muckle Flugga
2 Muckle Flugga sector old
3 Holm of Skaw
4 Balta Sound
5 Uyeasound
6 Bagi Stack
7 Whitehill
8 Ness of Sound
9 Muckle Holm
10 Little Holm
11 Tinga Skerry
12 Gruney
13 Point of Fethaland
14 Symbister Ness old
15 Ness of Queyfirth
16 Lamba
17 Little Roe
18 Skaw Taing Front
19 Mio Ness
20 Firths Voe
21 Lunna Holm
22 Muckle Skerry
23 Out Skerries
24 Grunay
25 Wether Holm
26 Suther Ness
27 Symbister Ness
28 Moul of Eswick
29 Hoo Stack
30 Rova Head
31 Twageos Point
32 Bressay
33 Mousa
34 Muckle Roe old
35 Sumburgh Head
36 Fugla Ness
37 Point of the Pund
38 North Havra
39 Vaila Sound
40 West Burrafirth Inner
41 West Burrafirth Outer
42 Housa Voe
43 Ve Skerries
44 Muckle Roe
45 Ness of Hillswick
46 Eshaness
47 Foula
48 Fair Isle North
49 Fair Isle South

Region 1

Shetland

1.1 MUCKLE FLUGGA (NORTH UNST) (i)

☆ *On Muckle Flugga island, North of Unst*

White tower (20 metres)
Established: 1854, re-built 1857
Designed by: Thomas and David Stevenson
Admiralty number: A3832
Light characteristic: Fl(2)W 20s
Grid reference: HP 607 197
Access inside tower: No

○ Visited on ....

Access to site: A chartered boat is required for a close view of the lighthouse, which is best seen from the sea or neighbouring island of Unst. On Unst, the lighthouse is visible from Hermaness headland, the most northerly point of Unst, which involves a 3.5kms (2 mile) walk on a reasonable path across open moorland. The lighthouse can also be seen on a clear day from a greater distance from the road beyond Saxa Vord village.

1.2 MUCKLE FLUGGA SECTOR (i)

Within Muckle Flugga lighthouse compound

○ Visited on ....

White square building
Light discontinued
Grid reference: HP 607 197
Access inside tower: No

Access to site: A chartered boat is required for a close view of the lighthouse, which is best seen from the sea. The building can be seen on a clear day from the road beyond Saxa Vord village on Unst.

Note: This sector light was displayed until a new light was introduced at the Holm of Skaw (see 1.3).

1.3 HOLM OF SKAW (i)

North east of Unst

○ Visited on ....

White metal framework tower (6 metres)
Admiralty number: A3834
Light characteristic: Fl W 5s
Grid reference: HP 669 171
Access inside tower: No

Access to site: Best seen from chartered boat. Otherwise the light is visible from a distance from the end of the road beyond Skaw on Unst.

1.4 BALTA SOUND (i)

South coast of Balta island

○ Visited on ....

White clad metal framework tower (5 metres)
Established: 2004
Admiralty number: A3828
Light characteristic: Fl WR 10s
Grid reference: HP 658 070
Access inside tower: No

Access to site: Best seen from chartered boat. Otherwise the light is visible from a distance from east of Ordale and Unst airport, which involves a walk of around 1km (0.6 miles) over rough ground from the end of the public road. On a clear day the lighthouse can be seen from the main A968 at Bobby's Bus Shelter on Unst.

1.5 UYEASOUND (i)

South coast of Unst

Metal framework tower on concrete structure (8 metres)
Admiralty number: A3830
Light characteristic: Fl(2)W 8s
Grid reference: HP 588 008
Access inside tower: No

○ Visited on ....

Access to site: From the end of the road to the south east of Uyeasound it is a short walk along the coast to the lighthouse.

1.6 BAGI STACK (i)

North west coast of Yell

○ Visited on ....

White clad metal framework tower (7 metres)
Established: 1976
Admiralty number: A3816
Light characteristic: Fl(4)W 20s
Grid reference: HP 478 050
Access inside tower: No

Access to site: Best seen from a chartered boat, as landing close to the lighthouse is very difficult. The light can be reached on foot by a walk of around 6kms (3.75 miles) across very rough and boggy terrain from Gloup, Yell. The light is otherwise very difficult to see.

1.7 WHITEHILL (VATSETTER) (i)

East coast of Yell

○ Visited on ....

White clad metal framework tower (5 metres)
Established: 2001
Admiralty number: A3824
Light characteristic: Fl WR 3s
Grid reference: HU 547 889
Access inside tower: No

Access to site: On a clear day the lighthouse is visible from the end of the public road at Vatsetter. A walk of around 2kms (1.25 miles) from here across flat, but boggy ground takes you to the lighthouse. The best view without the walk is by chartered boat. The lighthouse is visible from a distance on a clear day from the main road on Yell.

Note: For details of the previous lighthouse in this location see 6.37.

1.8 NESS OF SOUND (i)

South east coast of Yell

◯ Visited on ....

White metal framework tower (8 metres)
Admiralty number: A3814
Light characteristic: Fl(3)WRG 12s
Grid reference: HU 447 824
Access inside tower: No

Access to site: Accessible from the A968 just north of Clothan. From the main road it is a walk of around 1.5kms (1 mile) along a road and then rough ground to the lighthouse.

1.9 MUCKLE HOLM (i)

In Yell Sound

◯ Visited on ....

White clad metal framework tower (7 metres)
Established: 1976
Admiralty number: A3818
Light characteristic: Fl(4)W 10s
Grid reference: HU 403 888
Access inside tower: No

Access to site: Best seen by chartered boat. Otherwise, the light is visible from a distance from the A970 at North Roe, Shetland Mainland. Also visible from West Sandwick on Yell or, on a clear day, from the main road on Yell.

1.10 LITTLE HOLM (i)

In Yell Sound

◯ Visited on ....

White clad metal framework tower (7 metres)
Established: 1976
Admiralty number: A3819
Light characteristic: Iso W 4s
Grid reference: HU 404 862
Access inside tower: No

NESS OF SOUND

MUCKLE HOLM

Access to site: On a clear day the light is visible from a distance on minor roads north east of Housetter on Mainland Shetland, or visible from the east on the A968 on Yell. Best seen by chartered boat.

1.11 TINGA SKERRY (OUTER SKERRY) (i)

In Yell Sound

○ Visited on ....

White round tower (5 metres)
Admiralty number: A3815.6
Light characteristic: Q(2)G 10s
Grid reference: HU 414 807
Access inside tower: No

Access to site: Best seen by chartered boat. Alternatively, visible from a distance from Mio (see 1.19) Ness on Mainland Shetland, or from the main road on Yell or from Toft to Yell Ferry.

○ Visited on ....

1.12 GRUNEY (i)

North of Shetland Mainland

White tower (7 metres)
Established: 2004
Admiralty number: A3817
Light characteristic: Fl WR 5s
Grid reference: HU 382 968
Access inside tower: No

Access to site: Best seen by chartered boat. Landing on the island is difficult. The light is visible from Point of Fethaland (see 1.13).

GRUNEY

LITTLE HOLM

TINGA SKERRY

1.13 POINT OF FETHALAND (i) ✯

○ Visited on ....

Northern point of Shetland Mainland

White tower (7 metres)
Established: 1977
Admiralty number: A3817.5
Light characteristic: Fl(3)WR 15s
Grid reference: HU 378 947
Access inside tower: No

Access to site: Access is on foot from the end of the A970 at Isbister. The northerly route takes you 4.5kms (3 miles) across a clear track most of the way to the lighthouse.

1.14 SYMBISTER NESS OLD (i) ✯

○ Visited on ....

Relocated to private garden near Housetter, Mainland Shetland, from Whalsay

White metal tower with lantern
Established: 1904
Designed by: David A and Charles Stevenson
Light discontinued
Grid reference: HU 359 843
Access inside tower: No

Access to site: On private land, north east of the pier at Housetter. The lighthouse is visible from the main A970.

1.15 NESS OF QUEYFIRTH (i)

○ Visited on ....

North of Ollaberry

White tower (5 metres)
Admiralty number: A3820.5
Light characteristic: Oc WRG 6s
Grid reference: HU 371 825
Access inside tower: No

Access to site: Access is on foot, east-bound, from the end of the public road at Queyfirth. This walk is around 2kms (1.25 miles) across rough ground over Hoo Field hill.

1.16 LAMBA (i)

○ Visited on ....

In Yell Sound

White tower (8 metres)
Admiralty number: A3821
Light characteristic: Fl WRG 3s
Grid reference: HU 387 812
Access inside tower: No

Access to site: Best seen by chartered boat. Alternatively, the light is visible from the Ness of Queyfirth (see 1.15) or from the end of the public road at Queyfirth. Visible from Ollaberry on a clear day.

1.17 LITTLE ROE (i)

In Yell Sound

◯ Visited on ....

White GRP structure, orange band (5 metres)
Admiralty number: A3822.5
Light characteristic: Fl(3) WR 10s
Grid reference: HU 400 798
Access inside tower: No

Access to site: Best seen by chartered boat. Alternatively, the light is visible from the Ness of Queyfirth (see 1.15) or from the end of the public road at Queyfirth. Visible from Ollaberry on a clear day.

1.18 SKAW TAING FRONT (i)

○ Visited on ....

North of Sullom Voe

Orange and white GRP structure (5 metres)
Admiralty number: A3822.6
Light characteristic: Oc WRG 5s
Grid reference: HU 396 781
Access inside tower: No

Access to site: Access to the site is difficult due to its proximity to Sullom Voe. Best viewed by chartered boat or from a distance from Gluss Isle on Mainland Shetland. If accessing on foot, best approach is via Mio Ness (see 1.19), 4.8kms (3 miles) of rough terrain.

Note: The rear light is nearby.

1.19 MIO NESS (i)

○ Visited on ....

North east of Sullom Voe

White round tower (5 metres)
Admiralty number: A3815.3
Light characteristic: Q(2)WR 10s
Grid reference: HU 425 792
Access inside tower: No

Access to site: On foot from the road north of Toft by taking a gravel track north west towards a mast. From the mast head across rough ground for around 1.6km (1 mile). The lighthouse can be seen from a distance from Gluss Isle on Mainland Shetland.

1.20 FIRTHS VOE (i) ✭

○ Visited on ....

South of Toft

White tower (8 metres)
Established: 1909
Designed by: David A Stevenson
Admiralty number: A3812
Light characteristic: Oc WRG 8s
Grid reference: HU 454 747
Access inside tower: No

Access to site: Visible from a distance from main road south of Mossbank. The light can be accessed on foot from the main road by walking down a farm track.

1.21 LUNNA HOLM (i) ✯

○ Visited on ....

South of Yell

White round tower (8 metres)
Established: 1985
Admiralty number: A3811
Light characteristic: Fl(3)WRG 15s
Grid reference: HU 528 750
Access inside tower: No

Access to site: Best viewed by chartered boat. Otherwise, can be viewed from a distance from the main road on Yell or from Mossbank on Shetland Mainland.

1.22 MUCKLE SKERRY (i)

○ Visited on ....

North east of Whalsay island

White metal framework tower (5 metres)
Admiralty number: A3810
Light characteristic: Fl(2)WRG 10s
Grid reference: HU 626 734
Access inside tower: No

Access to site: The Vidlin to Out Skerries ferry takes you close to the lighthouse. On a clear day it can be seen from a distance from the golf club on Whalsay.

1.23 OUT SKERRIES (i) ✯

On Bound Skerry

◯ Visited on ....

White tower (30 metres)
Established: 1854, re-built 1857
Designed by: David and Thomas Stevenson
Admiralty number: A3807
Light characteristic: Fl W 20s
Grid reference: HU 701 719
Access inside tower: No

Access to site: Best viewed from chartered boat and landing is possible. Otherwise, can be seen from ferries to Out Skerries or from the neighbouring island of Bruray.

Note: There are plans for part of the lighthouse's original lens to be displayed at the Museum of Scottish Lighthouses, Fraserburgh, from 2019.

1.24 GRUNAY (i)

West coast of Grunay

◯ Visited on ....

White tower (2 metres)
Admiralty number: A3807.5
Light characteristic: Iso WRG 2s
Grid reference: HU 690 714
Access inside tower: No

Access to site: The lighthouse is visible from the south end of Bruray. There is a landing area on the island to which a chartered boat could be used.

1.25 WETHER HOLM (i)

West of Whalsay island

◯ Visited on

....

White metal framework tower (5 metres)
Established: 2006
Admiralty number: A3804.2
Light characteristic: Fl W 5s
Grid reference: HU 540 658
Access inside tower: No

Access to site: The light is visible from the Whalsay ferry from Vidlin. It can also be viewed from Suther Ness lighthouse on Whalsay (see 1.25).

1.26 SUTHER NESS (i)

West coast of Whalsay island

◯ Visited on ....

White metal framework tower (7 metres)
Admiralty number: A3804
Light characteristic: Fl WRG 3s
Grid reference: HU 551 654
Access inside tower: No

Access to site: The lighthouse is visible from the coastal road to the west of Brough on Whalsay. Walking to the lighthouse is possible from here.

Note: For details of the former Suther Ness lighthouse see 6.17.

1.27 SYMBISTER NESS (i)

South west end of Whalsay island

◯ Visited on ....

White metal framework tower (7 metres)
Admiralty number: A3802
Light characteristic: Fl(2)WG 12s
Grid reference: HU 532 622
Access inside tower: No

Access to site: From the harbour at Symbister, it is a short walk on grassy terrain from the end of the public road around the quarry to the lighthouse.

Note: For details of the former Symbister Ness lighthouse see 1.14.

1.28 MOUL OF ESWICK (i)

East of Eswick, Shetland Mainland

◯ Visited on ....

White clad metal framework tower (7 metres)
Admiralty number: A3796
Light characteristic: Fl WRG 3s
Grid reference: HU 500 535
Access inside tower: No

Access to site: The lighthouse is accessible on foot 1.5kms (1 mile) across rough and boggy terrain eastbound from Eswick.

1.29 HOO STACK (i)

North of Bressay

◯ Visited on

....

White clad metal framework tower (8 metres)
Admiralty number: A3795.7
Light characteristic: Fl(4)WRG 12s
Grid reference: HU 505 520
Access inside tower: No

Access to site: Best seen by chartered boat. The lighthouse is visible from a distance from Moul of Eswick lighthouse (see 1.27) or on a clear day from Bressay.

1.30 ROVA HEAD (i-t)

North of Lerwick, Shetland Mainland

◯ Visited on

....

White clad metal framework tower (7 metres)
Established: 2002
Admiralty number: A3794
Light characteristic: Fl(3)WRG 18s
Grid reference: HU 476 455
Access inside tower: No

Access to site: The lighthouse is on private land where access is restricted. It is, however, visible from the Lerwick to Out Skerries ferry.

1.31 TWAGEOS POINT (i)

In Lerwick harbour, Shetland Mainland

◯ Visited on

....

White round tower on square concrete base (2 metres)
Established: 1908
Admiralty number: A3778
Light characteristic: LFl W 6s
Grid reference: HU 483 408
Access inside tower: No

Access to site: The lighthouse is accessible on foot through a gate at the end of the houses on Twageos Road.

1.32 BRESSAY (i) ✯

○ Visited on ....

South west coast of Bressay island

White tower (16 metres)
Established: 1858
Designed by: David and Thomas Stevenson
Admiralty number: A3776
Light characteristic: Fl(2)W 20s
Grid reference: HU 489 376
Access inside tower: No

Access to site: Regular scheduled ferries operate to Bressay from Lerwick. The lighthouse is at the end of the public road beyond Kirkabister on Bressay.

Note: The former light apparatus from the lighthouse can now be seen in the Shetland Museum and Archives in Lerwick.

1.33 MOUSA (PEERIE BARD) (i-t)

○ Visited on ....

East of Mousa island

White clad metal framework tower (6 metres)
Admiralty number: A3772
Light characteristic: Fl W 3s
Grid reference: HU 471 239
Access inside tower: No

Access to site: Daily sailings (excluding Saturdays) operate from Sandwick to Mousa during the summer months. Peerie Bard island itself is tidal and access is prohibited for most of the year. The lighthouse can be seen from Mousa itself or from a distance from the coastal road at Mail, Mainland Shetland.

1.34 MUCKLE ROE OLD (i) ☆

○ Visited on ....

Relocated to side of access road to Sumburgh Head from location of current Muckle Roe light

White octagonal cast iron tower (8 metres)
Established: 1897
Designed by: David A and Charles Stevenson
Light discontinued
Grid reference: HU 409 083
Access inside tower: No

Access to site: Public parking for Sumburgh Head Visitor Centre is next to this lighthouse.

1.35 SUMBURGH HEAD (i) ☆

○ Visited on ....

Southern point of Shetland Mainland

White tower (17 metres)
Established: 1821
Designed by: Robert Stevenson
Admiralty number: A3766
Light characteristic: Fl(3)W 30s
Grid reference: HU 407 078

Access inside tower: No, although the Visitor Centre, open during the summer months, allows access to some of the associated buildings.

Access to site: There is a public car park around 1km from the lighthouse with an uphill walk on tarmac road. The former Muckle Roe lighthouse (see 1.34) can be seen close to the public car park.

FUGLA NESS

1.36 FUGLA NESS (i) ✯

North west coast of West Burra island, Shetland mainland

○ Visited on

....

White tower (8 metres)
Established: 1936
Admiralty number: A3858
Light characteristic: Fl(2)WRG 10s
Grid reference: HU 364 359
Access inside tower: No

Access to site: Accessible on foot from Hamnavoe. The lighthouse is visible from the village. A track between the houses takes you around the coast over grass and sections of large rocks.

Note: The old lens from the lighthouse is now on display at the Museum of Scottish Lighthouses, Fraserburgh.

1.37 POINT OF THE PUND (i)

West of Scalloway harbour, Shetland Mainland

○ Visited on

....

White GRP tower (3 metres)
Admiralty number: A3854.9
Light characteristic: Fl WRG 5s
Grid reference: HU 387 389
Access inside tower: No

Access to site: Access to the lighthouse is on foot through a clearly marked gate at the end of the public road in Port Arthur. From here it is a 1.5km (1 mile) walk on tarmac road and then rougher terrain.

1.38 NORTH HAVRA (i)

On North Havra island, south west of Tingwall

○ Visited on

....

White GRP tower (3 metres)
Admiralty number: A3854.7
Light characteristic: Fl(3)WRG 12s
Grid reference: HU 368 424
Access inside tower: No

Access to site: Best seen by chartered boat. The lighthouse is very difficult to see otherwise. Its light can be seen on a clear night from Scord of Sound.

POINT OF THE PUND

1.39 VAILA SOUND (WHITE NESS, RAM'S HEAD) (i)

○ Visited on ....

East of Vaila, Shetland Mainland

White clad metal framework tower (5 metres)
Admiralty number: A3850
Light characteristic: Fl WRG 8s
Grid reference: HU 246 462
Access inside tower: No

Access to site: Accessible on foot from the end of the public road at Vadlure. The walk on boggy, undulating terrain is around 2.5kms (1.5 miles) to the lighthouse.

1.40 WEST BURRAFIRTH INNER (i)

○ Visited on ....

North east of ferry terminal in West Burrafirth, Shetland Mainland

White square masonry tower (2 metres)
Admiralty number: A3845.5
Light characteristic: F WRG
Grid reference: HU 258 570
Access inside tower: No

Access to site: The lighthouse is accessible by a short grassy walk from the northbound turn-off in West Burrafirth. The light can also be viewed from the pier in West Burrafirth. The Outer light (see 1.41) can be seen from here.

1.41 WEST BURRAFIRTH OUTER (i)

○ Visited on ....

West of ferry terminal, West Burrafirth, Shetland Mainland

White square masonry tower (2 metres)
Admiralty number: A3844.9
Light characteristic: Oc WRG 8s
Grid reference: HU 244 570
Access inside tower: No

Access to site: The lighthouse is accessible by a short grassy walk northbound from the main road west of West Burrafirth. It can also be seen from the West Burrafirth Inner lighthouse (see 1.40) or from the ferry to Papa Stour.

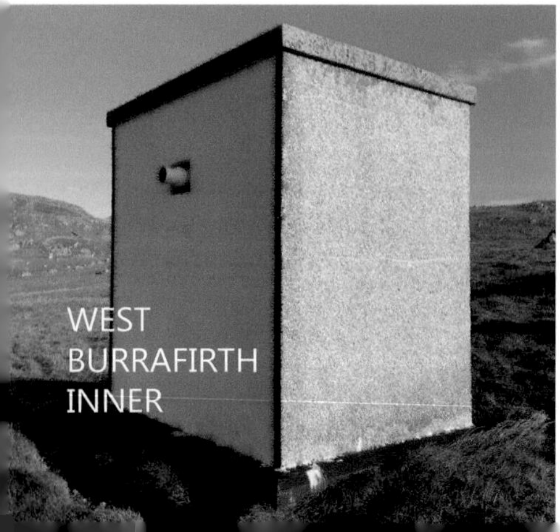

1.42 HOUSA VOE (i)

East coast of Papa Stour

○ Visited on ....

White concrete structure (2 metres)
Admiralty number: A3848
Light characteristic: Dir F WRG
Grid reference: HU 180 603
Access inside tower: No

Access to site: From the ferry terminal on Papa Stour it is a short walk around the coast to the lighthouse. It can be seen from a distance from the ferry on the approach to Papa Stour.

1.43 VE SKERRIES (i) ☆

North west of Papa Stour

○ Visited on ....

White concrete tower (16 metres)
Established: 1979
Admiralty number: A3847
Light characteristic: Fl(2)W 20s
Grid reference: HU 104 654
Access inside tower: No

Access to site: Best seen by chartered boat. On a clear day the lighthouse is visible from the north west coast of Papa Stour or from Eshaness lighthouse (see 1.46). On a particularly clear day the lighthouse can be seen from the A970 next to Loch of Haggrister.

1.44 MUCKLE ROE (i)

South coast of Muckle Roe

○ Visited on ....

White clad metal framework tower (7 metres)
Established: 2001
Admiralty number: A3844
Light characteristic: Fl WR 3s
Grid reference: HU 304 630
Access inside tower: No

Access to site: Access on foot from end of public road on Muckle Roe. A 2km (1.2 mile) partially signposted track takes you eastbound to the lighthouse along a rough, sloped path. The light is also visible from the road to Aith and Bixster.

Note: For details of the former Muckle Roe lighthouse see 1.34.

1.45 NESS OF HILLSWICK (i)

○ Visited on ....

South of Hillswick, Shetland Mainland

White clad metal framework tower (5 metres)
Admiralty number: A3840
Light characteristic: Fl(4)WR 15s
Grid reference: HU 278 745
Access inside tower: No

Access to site: Best seen by chartered boat. A walk from Hillswick of 2.5kms (1.5 miles) over rough ground takes you to the lighthouse. Alternatively, the lighthouse can be seen from a distance by walking from the end of the public road south west of Gunnister.

1.46 ESHANESS (i) ✯

○ Visited on ....

North west of Stenness, Shetland Mainland

White square tower (12 metres)
Established: 1929
Designed by: David A and Charles Stevenson
Admiralty number: A3838
Light characteristic: Fl W 12s
Grid reference: HU 206 784
Access inside tower: No

Access to site: From the B9078 west of Tangwick, take the westbound turning before you reach Stenness. There is a parking available next to the lighthouse.

1.47 FOULA (i) ✯

○ Visited on ....

South end of Foula island

White tower (8 metres)
Established: 1986
Admiralty number: A3860
Light characteristic: Fl(3)W 15s
Grid reference: HT 966 364
Access inside tower: No

Access to site: The island is served by plane or ferry depending on weather. Planes are available from Tingwall Airport on Mainland four times a week and it is possible to do a day trip to the island by this means. The scheduled passenger only ferry departs from Walls, West Mainland three times a week, meaning it is not possible to take a day trip to Foula using the ferry. From the airport the lighthouse is a 1.5km (1 mile) flat walk to the south along the road

and grassy terrain. From the ferry terminal it is a 3.5km (2 mile) walk to the west initially and then south, passing the airport, to reach the lighthouse.

1.48 FAIR ISLE NORTH (SKROO) (i) ☆

North east point of Fair Isle

○ Visited on ....

White tower (14 metres)
Established: 1892
Designed by: David A and Charles Stevenson
Admiralty number: A3756
Light characteristic: Fl(2)W 30s
Grid reference: HZ 222 741
Access inside tower: No

Access to site: Scheduled ferries and flights operate throughout the year, weather permitting. To reach the lighthouse it is a 4km (2.5 miles) walk on tarmac road from the airfield. It is 3km (1.8 miles) walk from the ferry terminal. The two lighthouses (see 1.49 for Fair Isle South) are 6.5kms (4 miles) apart if following the main road.

Note: There are plans for the original lens from the lighthouse to be displayed at the Museum of Scottish Lighthouses, Fraserburgh, from 2019.

1.49 FAIR ISLE SOUTH (SKADAN) (i) ☆

South west point of Fair Isle

○ Visited on ....

White tower (26 metres)
Established: 1892
Designed by: David A and Charles Stevenson
Admiralty number: A3750
Light characteristic: Fl(4)W 30s
Grid reference: HZ 198 698

Access inside tower: Tours are possible by prior arrangement.

Access to site: Scheduled ferries and flights operate throughout the year, weather permitting. To reach the lighthouse it is a 3.2km (2 mile) walk on tarmac road from the airfield. It is 4.5km (3 miles) walk from the ferry terminal. The two lighthouses (see 1.48 for Fair Isle North) are 6.5kms (4 miles) apart if following the main road.

Note: The former lens from the lighthouse in now on display in the Museum of Scottish Lighthouses, Fraserburgh.

Note: Fair Isle South was the last of the manned lighthouses operated by the Northern Lighthouse Board, with the last keepers leaving in 1998.

1 North Ronaldsay
2 North Ronaldsay old
3 Start Point
4 Noup Head
5 Calf of Eday
6 Papa Stronsay
7 Auskerry
8 Helliar Holm
9 Kirkwall West Pier
10 Copinsay
11 Rose Ness
12 Brough of Birsay
13 Hoy Low
14 Hoy High
15 Cava
16 Hoxa Head
17 Cantick Head
18 Tor Ness
19 Pentland Skerries High
20 Pentland Skerries Low
21 Sule Skerry

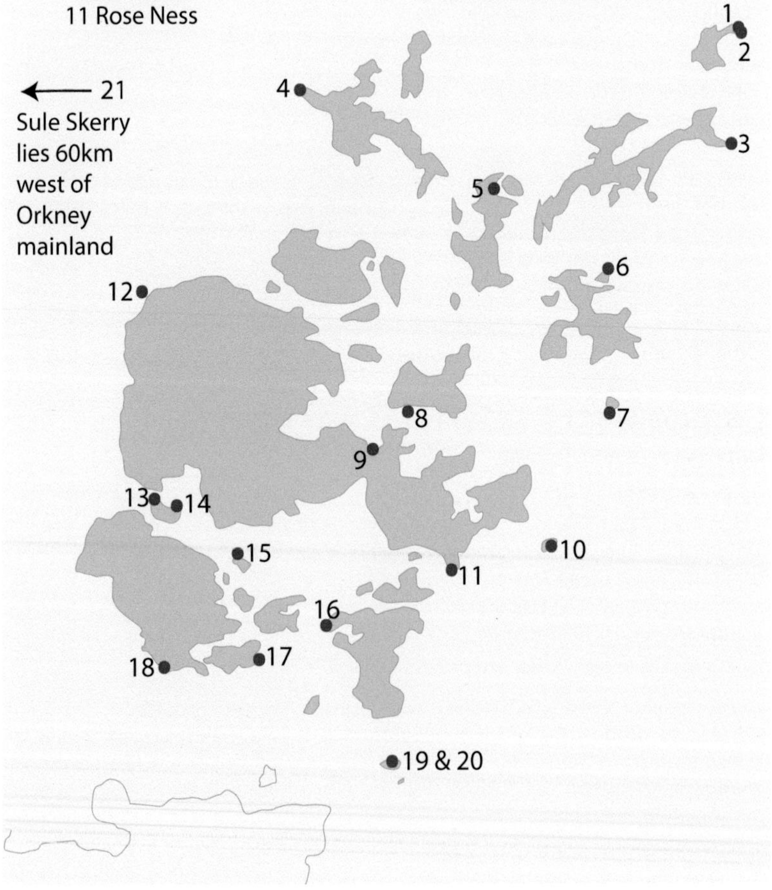

Region 2

Orkney

2.1 NORTH RONALDSAY (i) ☆

Northern point of North Ronaldsay island

Red brick tower, 2 white bands (42 metres)
Established: 1854
Designed by: Alan Stevenson
Admiralty number: A3722
Light characteristic: Fl W 10s
Grid reference: HY 784 559
Access inside tower: The Visitor Centre includes guided tours of the lighthouse.

○ Visited on ....

Access to site: North Ronaldsay can be reached by plane or vehicle ferry from Kirkwall, or ferry from Papa Westray. Once on the island, it is 5kms (3 miles) from the airport, or 6.5kms (4 miles) from the ferry terminal. Parking is available at the lighthouse. The old lighthouse (see 2.2) is a short walk away. An overnight or longer stay on the island is recommended to avoid hurrying the island and experience.

2.2 NORTH RONALDSAY OLD (DENNIS HEAD OLD BEACON) (i)

○ Visited on ....

Northern point of North Ronaldsay island

Round stone tower (21 metres)
Established: 1789
Designed by: Thomas Smith
Light discontinued
Grid reference: HY 790 553
Access inside tower: No

Access to site: North Ronaldsay can be reached by plane or vehicle ferry from Kirkwall, or ferry from Papa Westray. Once on the island, it is 5kms (3 miles) from the airport, or 6.5kms (4 miles) from the ferry terminal. Parking is available at the operational lighthouse (see 2.1) from which it is a short walk across fields. An overnight or longer stay on the island is recommended to avoid hurrying the island and experience.

2.3 START POINT (i-t) ✭

○ Visited on ....

North eastern point of Sanday island

White tower, black stripes (23 metres)
Established: 1806
Designed by: David and Thomas Stevenson
Admiralty number: A3718
Light characteristic: Fl(2)W 20s
Grid reference: HY 787 434
Access to tower: Occasional tours of the tower are run by the local ranger. See sandayranger.com for details.

Access to site: The lighthouse is accessible at low tide from Sanday, which is accessible by scheduled ferry from Kirkwall. It should be noted that it may not

be possible to visit the lighthouse on a day trip to Sanday, so an overnight stay on the island is recommended. There is a small parking area at the end of the public single-track road to the east of Northwall. A bus is available from the ferry in the summer months. It is a 3.75km (2.25 mile) return walk along a rough track including slippery rocks on the tidal section.

2.4 NOUP HEAD (i) ☆

West coast of Westray island

○ Visited on ....

White tower (24 metres)
Established: 1898
Designed by: David A and Charles Stevenson
Admiralty number: A3736
Light characteristic: Fl W 30s
Grid reference: HY 392 499
Access inside tower: No

Access to site: Westray is accessible by scheduled vehicle and passenger ferry from Kirkwall. The lighthouse is at the end of the road west of Pierowall. Although there is parking available next to the lighthouse the last 1.6kms (1 mile) of the journey is on very rough ground and may not be suitable for all vehicles.

2.5 CALF OF EDAY (CALFSOUND) (i) ☆

North east coast of Eday island

○ Visited on ....

White tower (9 metres)
Established: 2002
Admiralty number: A3728
Light characteristic: Fl(3)WRG 10s
Grid reference: HY 566 392

Access inside tower: No

Access to site: Scheduled vehicle ferry crossings operate to Eday island from Kirkwall. From the ferry terminal it is a 9km (5.4 mile) drive, walk or taxi ride to Calfsound. Parking is available to the east of the beach, followed by a 1.5km (1 mile) walk along the beach and across fields.

2.6 PAPA STRONSAY (i)

North east coast Papa Stronsay island, north of Stronsay

○ Visited on ....

White metal framework tower (5 metres)
Established: 2002
Admiralty number: A3706
Light characteristic: Fl(4)W 20s
Grid reference: HY 668 300
Access inside tower: No

Access to site: The island is privately owned and run as a monastery. The best views are from a chartered boat.

2.7 AUSKERRY (i) ☆

Southern point of Auskerry island

○ Visited on ....

White tower (34 metres)
Established: 1866
Designed by: David and Thomas Stevenson
Admiralty number: A3680
Light characteristic: Fl W 20s
Grid reference: HY 672 155
Access inside tower: No

Access to site: Best seen by chartered boat. It may be possible to organise a visit to the island with the owners during the summer months. Contact details are available at isleofauskerry.com. On a clear day the lighthouse can be seen from a distance from sailings between Kirkwall and Lerwick.

2.8 HELLIAR HOLM (i) ✯

South coast of Shapinsay island

○ Visited on ....

White tower (13 metres)
Established: 1893
Designed by: David A and Charles Stevenson
Admiralty number: A3684
Light characteristic: Fl WRG 10s
Grid reference: HY 484 150
Access inside tower: No

Access to site: The lighthouse can be seen from ferries between Kirkwall and Lerwick or Aberdeen. It is also possible to see the lighthouse from Work to the north east of Kirkwall. A closer view would require a chartered boat.

Note: The former lens from Helliar Holm lighthouse is now on display at the Northern Lighthouse Board headquarters in Edinburgh, which is open to the public on special occasions.

2.9 KIRKWALL WEST PIER (i) ✯

On west pier in Kirkwall harbour

○ Visited on ....

White round cast iron tower with ornate lantern (approx. 6 metres)
Established: 1854
Light discontinued
Grid reference: HY 449 113
Access inside tower: No

Access to site: There is public access to the pier from Harbour Street. A public car park is next to the pier entrance.

2.10 COPINSAY (i) ✯

East coast of Copinsay island

○ Visited on ....

White tower (16 metres)

HELLIAR HOLM

KIRKWALL WEST PIER

Established: 1915
Designed by: David A and Charles Stevenson
Admiralty number: A3676
Light characteristic: Fl(5)W 30s
Grid reference: HY 614 012
Access inside tower: No

Access to site: Best seen by chartered boat. On a clear day the lighthouse is visible from Point of Ayre in Deerness.

2.11 ROSE NESS (i) ☆

○ Visited on ....

South coast of Orkney Mainland

White tower (7 metres)
Admiralty number: A3626
Light characteristic: Fl W 6s
Grid reference: ND 522 987
Access inside tower: No

Access to site: A gate at the end of the public Upper Cornquoy Road takes you to the coast. A walk of around 2kms (1.25 miles) from the gate across grass takes you to the lighthouse.

Note: The lantern and lens from the previous light in this location is now on display at the Museum of Scottish Lighthouses, Fraserburgh.

2.12 BROUGH OF BIRSAY (i-t) ✯

○ Visited on ....

North west coast of Birsay island

White castellated tower and building (11 metres)
Established: 1925
Designed by: David A and Charles Stevenson
Admiralty number: A3700
Light characteristic: Fl(3)W 25s
Grid reference: HY 235 285
Access inside tower: No

Access to site: The lighthouse is accessible at low tide across a causeway from Birsay, where there is parking, followed by a 1km (0.6 mile) slight uphill walk on grass.

2.13 HOY MOUTH FRONT ✯ (HOY LOW) (i)

○ Visited on ....

North west coast of Graemsay island

Sandstone tower (12 metres)
Established: 1851
Designed by: Alan Stevenson
Admiralty number: A3644
Light characteristic: Iso W 3s
Grid reference: HY 247 066
Access inside tower: No

Access to site: Scheduled (passenger only) ferries operate to Graemsay from Stromness. From the ferry terminal it is a 4km (2.3 mile) walk along a tarmac road and a short section of rough track on final approach. The route also takes in the Rear lighthouse (see 2.14). The tower is also visible from the Scrabster ferry on the approach to Stromness.

Note: The old lens from the lighthouse is now on display in the Stromness Museum.

2.14 HOY MOUTH REAR (HOY HIGH) (i) ☆

○ Visited on ....

North east coast of Graemsay island

White tower (33 metres)
Established: 1851
Designed by: Alan Stevenson
Admiralty number: A3644.1
Light characteristic: Oc WR 8s
Grid reference: HY 268 060
Access inside tower: No

Access to site: The buildings are a private home, but the tower is visible from the road. Scheduled (passenger only) ferries operate to Graemsay from Stromness. A 1km (0.7 mile) walk along a tarmac road takes you to the lighthouse. Continuing a further 3km (1.8 mile) walk takes you to the Front light (see 2.13). The tower is also visible from the Scrabster ferry on the approach to Stromness.

2.15 CAVA (i) ☆

○ Visited on ....

Northern point of Cava island

White round tower (8 metres)
Admiralty number: A3636
Light characteristic: Fl WR 3s
Grid reference: HY 322 006
Access inside tower: No
Access to site: Best seen by chartered boat. The lighthouse is clearly visible from the Houton ferry to Lyness on Hoy.

Notes: It is possible to see the Barrel of Butter light, which is an interesting structure but does not qualify for its own listing in this book, on a clear day from Cava lighthouse and the ferry to Hoy.

2.16 HOXA HEAD (i)

○ Visited on ....

North west coast of South Ronaldsay island

White clad metal framework tower (7 metres)
Established: 1996
Admiralty number: A3610
Light characteristic: Fl WR 3s
Grid reference: ND 404 932
Access inside tower: No

Access to site: Public parking is available beyond Hoxa and Uppertown at the end of the public road. From here it is less than a 1km (0.6 mile) walk on a road and track which takes you to the lighthouse.

Note: For details of the former lighthouse in this location see 6.18.

2.17 CANTICK HEAD (i) ✯

○ Visited on ....

East coast of South Walls island

White tower (22 metres)
Established: 1858
Designed by: David and Thomas Stevenson
Admiralty number: A3602
Light characteristic: Fl W 20s
Grid reference: ND 347 894
Access inside tower: No

Access to site: Hoy is accessible by scheduled ferry from Houton on Orkney Mainland. From Lyness on Hoy, South Walls island and the lighthouse are accessible by road across a causeway on the B9047. The distance by road from Lyness to the lighthouse is 17kms (10.6 miles). There is a small area to park outside the lighthouse compound.

Note: The old light mechanism from the lighthouse can now been seen at the Scapa Flow Visitor Centre and Museum in Lyness.

2.18 TOR NESS (i) ☆

South coast of Hoy

◯ Visited on

....

White tower (8 metres)
Admiralty number: A3600
Light characteristic: Fl W 5s
Grid reference: ND 251 886
Access inside tower: No

Access to site: Hoy is accessible by scheduled ferry from Houton on Orkney Mainland. It is 9kms (5.6 miles) from the ferry terminal to North Bay beach. From here it is a further 3km (1.8 mile) walk to the lighthouse, passing Melsetter Farm.

2.19 PENTLAND SKERRIES HIGH (i) ☆

North coast of Muckle Skerry, Pentland Skerries

◯ Visited on

....

White tower (36 metres)
Established: 1794
Designed by: Thomas Smith and Robert Stevenson
Admiralty number: A3562
Light characteristic: Fl(3)W 30s
Grid reference: ND 465 784
Access to tower: No

Access to site: Occasional organised boat trips operate out of Wick, although landing on the island is not permitted. The low light (see 2.20) is nearby.

2.20 PENTLAND SKERRIES LOW (i)

○ Visited on ....

Shorter tower to the south of operational light

White tower (18 metres)
Established: 1794
Designed by: Thomas Smith and Robert Stevenson
Light discontinued 1895
Grid reference: ND 465 783
Access to tower: No

Access to site: Occasional organised boat trips operate out of Wick, although landing on the island is not permitted. The high light (see 2.19) is nearby.

2.21 SULE SKERRY (i) ☆

○ Visited on ....

On north coast of Sule Skerry island, 60kms (40 miles) west of Orkney

White tower (27 metres)
Established: 1895
Designed by: David A and Charles Stevenson
Admiralty number: A3868
Light characteristic: Fl(2)W 15s
Grid reference: HX 624 246
Access inside tower: No

Access to site: The lighthouse is only visible by chartered boat. Landing on the island is challenging and requires extremely calm conditions.

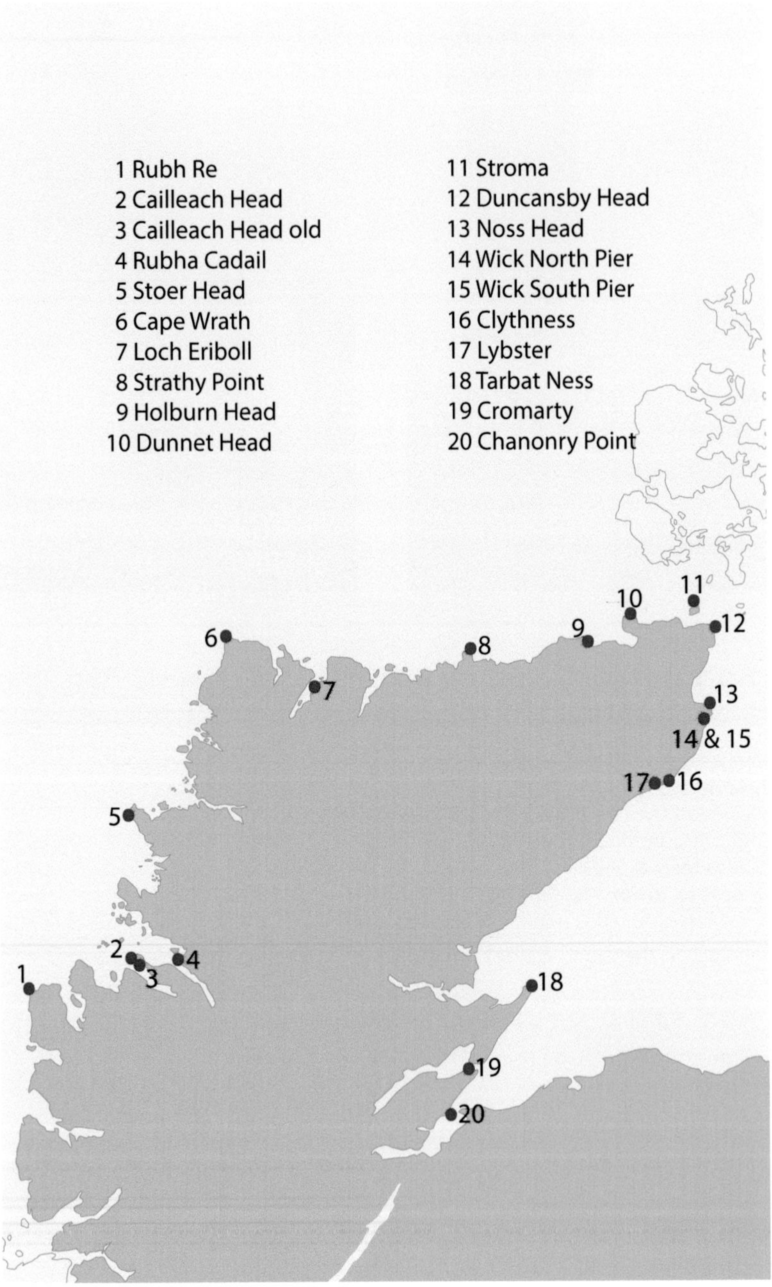
1 Rubh Re
2 Cailleach Head
3 Cailleach Head old
4 Rubha Cadail
5 Stoer Head
6 Cape Wrath
7 Loch Eriboll
8 Strathy Point
9 Holburn Head
10 Dunnet Head
11 Stroma
12 Duncansby Head
13 Noss Head
14 Wick North Pier
15 Wick South Pier
16 Clythness
17 Lybster
18 Tarbat Ness
19 Cromarty
20 Chanonry Point
1
2
3
4
5
6
7
8
9
10
11
12
13
14 & 15
16
17
18
19
20

Region 3

Northern Scotland

White tower (25 metres)

3.1 RUBH RE (RUBHA REIDH) (m) ☆

North of Melvaig

Established: 1912
Designed by: David A and Charles Stevenson
Admiralty number: A3900
Light characteristic: Fl(4)W 15s
Grid reference: NG 740 919
Access inside tower: No

○ Visited on ....

Access to site: The lighthouse is accessible on foot along the 5km (3 mile) road from Melvaig. It is advisable to check before travelling if you are planning to drive to the lighthouse as restrictions may be in place for those not staying at the lighthouse.

Note: The old lens from the lighthouse can now been seen at the Gairloch Heritage Museum.

3.2 CAILLEACH HEAD (m)

○ Visited on ....

At end of Scoraig peninsula

White clad metal framework tower (5 metres)
Established: 2001
Admiralty number: A3887
Light characteristic: Fl(2)W 12s
Grid reference: NG 986 985
Access inside tower: No

Access to site: Accessible by local boat from Badluarach at high tide followed by a long walk. Contact information for local boats are available on the Scoraig Community Association website. Alternative access to Scoraig village is a 16.5km (10.25 mile) walk from Badluarach on a good, but often narrow path. From the jetty head west and a 3km (1.8 mile) walk on rough terrain with no clear path takes you to the lighthouse. The light can be seen from a distance from the Ullapool to Stornoway ferry.

Note: For details of the former lighthouse in this location see 3.3.

3.3 CAILLEACH HEAD OLD (m) ✯

○ Visited on ....

Relocated to Scoraig village from Cailleach Head

White octagonal tower (4 metres)
Established: 1953
Light discontinued 2001
Grid reference: NH 009 965
Access inside tower: Yes. The old lighthouse contains a museum about life in Scoraig.

Access to site: Easiest route is by local boat from Badluarach at high tide. Contact information for local boats are available on the Scoraig Community Association website. From the jetty follow the coast past east, continuing as it veers north and then east. After a short distance the lighthouse is on the left just beyond a section of trees. The walk from the jetty to the lighthouse is around 1.5kms (1 mile). Alternative access to Scoraig is a 16.5km (10.25 mile) walk from Badluarach on a good, but often narrow path.

3.4 RUBHA CADAIL (m) ✯

West of Rhue

○ Visited on

....

White tower (9 metres)
Established: 1952
Admiralty number: A3885
Light characteristic: Fl WRG 6s
Grid reference: NH 092 974
Access inside tower: No

Access to site: To the west of Rhue village public parking is available. The lighthouse is visible from the car park and it is a short walk across grass to reach the tower.

3.5 STOER HEAD (m) ✯

North west of the village of Stoer

○ Visited on

....

White tower (14 metres)
Established: 1870
Designed by: David and Thomas Stevenson
Admiralty number: A3882
Light characteristic: Fl W 15s
Grid reference: NC 004 330
Access inside tower: No

Access to site: Point of Stoer is signposted from the B869 to the north of Lochinver. Once at Stoer Head, it is a short walk on a tarmac road from the public parking area to the lighthouse.

RUBHA CADAIL

3.6 CAPE WRATH (m) ✯

○ Visited on ....

North west point of mainland

White tower (20 metres)
Established: 1828
Designed by: Robert Stevenson
Admiralty number: A3880
Light characteristic: Fl(4)W 30s
Grid reference: NC 260 748
Access inside tower: No

Access to site: Cape Wrath is a military training area, and access is not possible when training operations are underway. It is advisable to check before travelling. Easiest access is using regular (seasonal) passenger only boat crossings from Keoldale and the minibus to the lighthouse. A considerably more challenging route is on foot from Sandwood Bay, over 12km (7.5 miles) to the south on very rough terrain with no clear path.

Note: The old lens from the lighthouse is due for relocation to Sumburgh Head Visitor Centre in 2019.

3.7 LOCH ERIBOLL (WHITE HEAD) (m)

○ Visited on ....

East coast of Loch Eriboll

White metal framework tower (5 metres)
Established: 2003
Admiralty number: A3874
Light characteristic: Fl WR 10s
Grid reference: NC 459 618
Access inside tower: No

Access to site: The lighthouse is a 1.5km (1 mile) walk due north across open, sometime boggy, moorland with no clear path. The best starting point is from the main A836 just before it meets the east side of Loch Eriboll.

3.8 STRATHY POINT (m) ✯

Northern point of Strathy

○ Visited on ....

White tower (14 metres)
Established: 1958
Designed by: Peter Hyslop
Light discontinued 2012
Grid reference: NC 828 697
Access inside tower: No

Access to site: There is a car parking area at the end of the public road towards Strathy Point. The lighthouse can be reached by walking 1km (0.6 miles) on a tarmac road from here.

3.9 HOLBURN HEAD (m) ✯

East end of Scrabster harbour

○ Visited on ....

White tower (17 metres)
Established: 1862
Designed by: David and Thomas Stevenson
Light discontinued 2003
Grid reference: ND 107 707
Access inside tower: No

Access to site: A short walk from Scrabster harbour on a tarmac road takes you to the lighthouse, which is now a privately-owned property.

3.10 DUNNET HEAD (m) ✯

North of Dunnet

○ Visited on ....

White tower (20 metres)
Established: 1831
Designed by: Robert Stevenson

Admiralty number: A3574
Light characteristic: Fl(4)W 30s
Grid reference: ND 203 768
Access inside tower: No

Access to site: The lighthouse is signposted from the A836, the main road through Dunnet village. There is a large public parking area next to the lighthouse compound, which is privately owned, but occasionally open to the public.

Note: The old lens from the lighthouse is now on display at the Museum of Scottish Lighthouses, Fraserburgh.

3.11 STROMA (i) ☆

○ Visited on ....

North coast of Stroma island

White tower (23 metres)
Established: 1896
Designed by: David A and Charles Stevenson
Admiralty number: A3568
Light characteristic: Fl(2)W 20s
Grid reference: ND 354 792
Access inside tower: No

Access to site: Access to the island is via boat charter with the island's owner only from Gills Bay. Once on the island, a 3km (1.8 mile) walk on clear path/road takes you to the lighthouse.

3.12 DUNCANSBY HEAD (m) ☆

○ Visited on ....

East of John O'Groats

White square tower (11 metres)
Established: 1924
Designed by: David A and Charles Stevenson

Admiralty number: A3558
Light characteristic: Fl W 12s
Grid reference: ND 406 733
Access inside tower: No

Access to site: Duncansby Head is signposted from the A99, the main road into John O'Groats. Public parking is available close to the lighthouse.

3.13 NOSS HEAD (m) ☆

○ Visited on ....

North of Wick

White tower (18 metres)
Established: 1849
Designed by: Alan Stevenson
Admiralty number: A3544
Light characteristic: Fl WR 20s

Grid reference: ND 389 551
Access inside tower: No

Access to site: Follow signs to Staxigoe from Wick town centre and then a sign for Noss Head. A short walk on a tarmac road from the public car park takes you to the lighthouse.

Note: The old lens from Noss Head lighthouse can now been seen at the nearby Wick Heritage Centre.

3.14 WICK NORTH PIER (m)

○ Visited on ....

Towards end of north pier in Wick harbour

Hexagonal iron tower with lantern and gallery (12 metres)
Established: 1910
Light discontinued
Grid reference: ND 368 507
Access inside tower: No

Access to site: The lighthouse is located within an operational harbour but can be viewed from the south side of the harbour. The lighthouse on the south pier (see 3.15) is nearby.

3.15 WICK SOUTH PIER (m) ☆

○ Visited on ....

On end of south pier in Wick harbour

White octagonal tower (11 metres)
Established: 1897
Admiralty number: A3528
Light characteristic: Fl WRG 3s
Grid reference: ND 371 506
Access inside tower: No

Access to site: The lighthouse is accessible along the south pier from the main harbour road. The light on the north pier (see 3.14) is nearby.

3.16 CLYTHNESS (m) ✯

○ Visited on ....

North of Lybster

White round tower, red band (13 metres)
Established: 1916
Designed by: David A and Charles Stevenson
Light discontinued 2010
Grid reference: ND 291 364
Access inside tower: No

Access to site: The tower and buildings are now privately owned. Parking is available at a lay-by a short distance to the north of the lighthouse on the A99.

3.17 LYBSTER SOUTH PIER (m) ✯

○ Visited on ....

In Lybster harbour

White tower (approx. 10 metres)
Established: 1884
Admiralty number: A3522
Light characteristic: Oc R 6s
Grid reference: ND 245 348
Access inside tower: No

Access to site: Parking is available at the Waterlines Heritage Centre. From the car park it is a short walk to the pier.

3.18 TARBAT NESS (m) ✯

○ Visited on ....

East of Tain

White tower, red bands (41 metres)
Established: 1830
Designed by: Robert Stevenson
Admiralty number: A3506

Light characteristic: Fl(4)W 30s
Grid reference: NH 947 876
Access inside tower: No

Access to site: Access is via the Tarbatness Road turn-off, just after entering the village of Portmahomack, and then Golf Course Road. The lighthouse is a short walk on a tarmac road from the public car park.

Note: The old lens from the lighthouse is now on display at the National Maritime Museum in Greenwich, London.

3.19 CROMARTY (m) ☆

North east coast of the Black Isle

○ Visited on ....

White tower (13 metres)
Established: 1846
Designed by: Alan Stevenson
Light discontinued: 2006
Grid reference: NH 787 678

Access inside tower: The lighthouse is occasionally open for special events but is not routinely accessible.

Access to site: The site is easily accessible along George Street in Cromarty.

3.20 CHANONRY POINT (m) ✯

○ Visited on

....

South east coast of the Black Isle

White tower (13 metres)
Established: 1846
Designed by: Alan Stevenson
Admiralty number: A3440
Light characteristic: Oc W 6s
Grid reference: NH 750 557
Access inside tower: No

Access to site: Public parking is available at end of Ness Road, Fortrose. Chanonry Point is signposted from the A832, High Street.

Note: The former lens from the lighthouse is now on display in the Museum of Scottish Lighthouses, Fraserburgh.

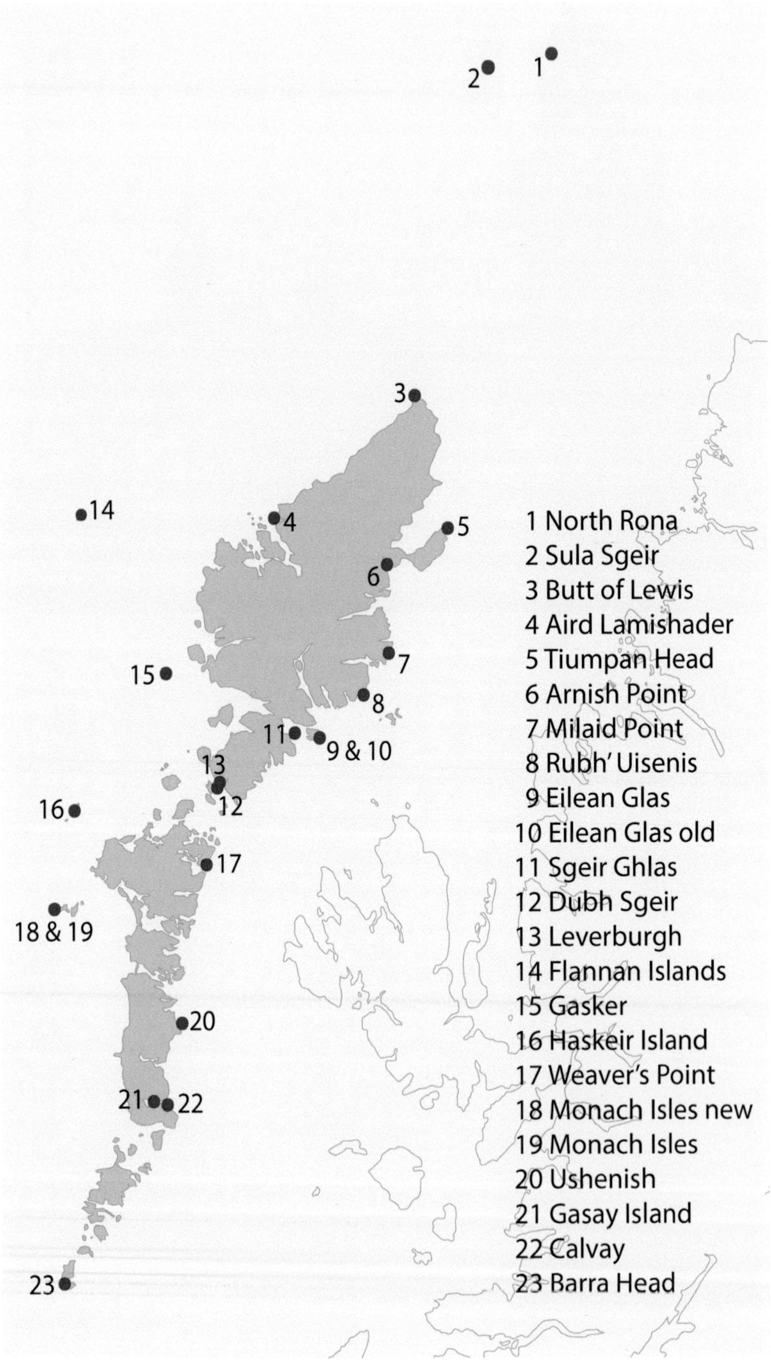

1
2
3
4
5
6
7
8
9 & 10
11
12
13
14
15
16
17
18 & 19
20
21
22
23
1 North Rona
2 Sula Sgeir
3 Butt of Lewis
4 Aird Lamishader
5 Tiumpan Head
6 Arnish Point
7 Milaid Point
8 Rubh' Uisenis
9 Eilean Glas
10 Eilean Glas old
11 Sgeir Ghlas
12 Dubh Sgeir
13 Leverburgh
14 Flannan Islands
15 Gasker
16 Haskeir Island
17 Weaver's Point
18 Monach Isles new
19 Monach Isles
20 Ushenish
21 Gasay Island
22 Calvay
23 Barra Head

Region 4

Western Isles

4.1 NORTH RONA (i) ☆

On east coast of North Rona

White square tower (9 metres)
Established: 1984
Designed by: J H K Williamson
Admiralty number: A3869
Light characteristic: Fl(3)W 20s
Grid reference: HW 818 323
Access inside tower: No

○ Visited on

....

Access to site: The lighthouse is visible from a chartered boat only. Landing on the island is possible in calm conditions.

4.2 SULA SGEIR (i)

West of North Rona

○ Visited on ....

Square metal structure (5 metres)
Admiralty number: A3870
Light characteristic: Fl W 15s
Grid reference: HW 619 304
Access inside tower: No

Access to site: The lighthouse is visible from a chartered boat only. Landing on the island is possible in calm conditions, but restrictions are in place during the summer months while the birds are nesting.

4.3 BUTT OF LEWIS (i) ☆

Northern point of Isle of Lewis

○ Visited on ....

Red brick tower (37 metres)
Established: 1862
Designed by: David and Thomas Stevenson
Light characteristic: Fl W 5s
Grid reference: NB 520 665
Access inside tower: No

Access to site: The lighthouse is accessible by road heading north east from the village of Eoropie. There is public parking available next to the lighthouse compound.

Note: The old lens from the lighthouse is currently in storage at the Museum of Scottish Lighthouses, Fraserburgh.

4.4 AIRD LAMISHADER (i)

North east entrance to Loch Roag, Isle of Lewis

○ Visited on ....

White clad metal framework tower
Admiralty number: A4032
Light characteristic: Fl W 6s
Grid reference: NB 173 430
Access inside tower: No

Access to site: A 2km (1.25 mile) walk on rough and rocky terrain with no clear path westbound from Borrowston (Borghastan) village takes you to the lighthouse.

BUTT OF LEWIS

4.5 TIUMPAN HEAD (i) ✯

○ Visited on ....

Northern point of Eye Peninsula, Isle of Lewis

White tower (21 metres)
Established: 1900
Designed by: David A and Charles Stevenson
Admiralty number: A3972
Light characteristic: Fl(2)W 15s
Grid reference: NB 574 377
Access inside tower: No

Access to site: The lighthouse is accessible by taking the single-track road north from the village of Portvoller. A small parking area is available close to the lighthouse compound.

4.6 ARNISH POINT (i) ✯

○ Visited on ....

West of entrance to Stornoway harbour, Isle of Lewis

White tower (14 metres)
Established: 1853
Designed by: Alan Stevenson
Admiralty number: A3976
Light characteristic: Fl WR 10s
Grid reference: NB 433 308
Access inside tower: No

Access to site: Only access to the tower is through the private Arnish Point Industrial Estate. The lighthouse is visible from the Ullapool ferry on the approach to Stornoway.

4.7 MILAID POINT (GOB NA MILAID) (i)

○ Visited on ....

East of Leumrabhagh, east coast of Isle of Lewis

White clad metal framework tower
Admiralty number: A3984
Light characteristic: Fl W 15s

TIUMPAN HEAD

ARNISH POINT

Grid reference: NB 422 115
Access inside tower: No

Access to site: The light is best viewed by boat. Access by land is very challenging across 6kms (3.75 miles) of difficult, uneven and boggy terrain from the main road north of the junction for Orasaigh, before Leumrabhagh.

4.8 RUBH' UISENIS (i) ✯

South east coast of Isle of Lewis

○ Visited on ....

White tower (7 metres)
Established: 1885
Admiralty number: A3986
Light characteristic: Fl W 5s
Grid reference: NB 354 030
Access inside tower: No

Access to site: The light is best viewed by boat. Access by land is very challenging across 15kms (9 miles) of extremely difficult terrain with no clear path from the end of the road at Eisgein.

4.9 EILEAN GLAS (i) ✯

Eastern point of Scalpay, Isle of Harris

○ Visited on ....

White tower, red bands (30 metres)
Established: 1824
Designed by: Robert Stevenson
Admiralty number: A3990
Light characteristic: Fl(3)W 20s
Grid reference: NG 247 947
Access inside tower: No

Access to site: The lighthouse is accessible through the gate at Kennavay, from which it is a 2km (1.25 mile) walk on a tarmac road and path. The old lighthouse (see 4.10) is also within the compound.

Note: The old lens from the lighthouse is now on display at the Science Museum in London.

4.10 EILEAN GLAS OLD (i)

○ Visited on ....

Next to current Eilean Glas lighthouse

White tower, no lantern (6.5 metres)
Established: 1789
Designed by: Thomas Smith
Light discontinued 1824
Grid reference: NG 247 947
Access inside tower: Access to ground floor level only.

Access to site: The lighthouse is accessible through the gate at Kennavay, from which it is a 2km (1.25 mile) walk on a tarmac road and path. The operational lighthouse (see 4.9) is also within the compound.

4.11 SGEIR GHLAS (i) ☆

○ Visited on ....

South east of Tarbert, Isle of Harris

White round concrete tower (5 metres)
Admiralty number: A3993
Light characteristic: Iso WRG 4s
Grid reference: NG 182 970
Access inside tower: No

Access to site: Access to the island is by chartered boat only. The lighthouse is visible from the Uig ferry on the approach to Tarbert.

4.12 DUBH SGEIR (i)

○ Visited on ....

South west of Leverburgh

Red concrete tower, black bands (6 metres)
Admiralty number: A3994.2
Light characteristic: Q(2)W 5s
Grid reference: NG 001 855
Access inside tower: No

Access to site: Access to the island by chartered boat, although the light is visible from scheduled boats leaving or approaching Leverburgh.

LEVERBURGH

FLANNAN ISLANDS

4.13 LEVERBURGH (i)

○ Visited on ....

North west of Leverburgh pier

White octagonal concrete tower, no lantern (9 metres)
Light discontinued
Grid reference: NG 009 867
Access inside tower: No

Access to site: The lighthouse is accessible by taking a 1km (0.6 mile) walk over moderate terrain from the main A859 road. The best starting point is the Glen Kyles area between Leverburgh and Northton.

4.14 FLANNAN ISLANDS (EILEAN MÒR) (i) ☆

○ Visited on ....

North coast of Eilean Mòr island, Flannan Isles

White tower (23 metres)
Established: 1899
Designed by: David A and Charles Stevenson
Admiralty number: A4028
Light characteristic: Fl(2)W 30s
Grid reference: NA 727 469
Access inside tower: No

Access to site: The lighthouse is accessible by chartered boat only. Landing on the island requires great care and very calm sea conditions.

Note: A small remaining piece of the old destroyed lens is now on display at the Museum of Scottish Lighthouses, Fraserburgh.

SGEIR GHLAS

4.15 GASKER (i)

○ Visited on ....

South west of Scarp

White clad metal framework tower (5 metres)
Admiralty number: A4030
Light characteristic: Fl(3)W 10s
Grid reference: NA 876 117
Access inside tower: No

Access to site: By chartered boat only. It is very challenging landing on the island, only possible in very calm conditions. However, the light is clearly visible without the need to land.

4.16 HASKEIR ISLAND (i) ☆

○ Visited on ....

North west of North Uist

White glass reinforced plastic tower (9 metres)
Established: 1997
Admiralty number: A4020.3
Light characteristic: Fl W 20s
Grid reference: NF 613 820
Access inside tower: No

Access to site: By chartered boat only. Landing is only possible in calm conditions and reaching the lighthouse requires a short section of scrambling on rocks.

4.17 WEAVER'S POINT (i)

○ Visited on ....

North of entrance to Lochmaddy, North Uist

White clad metal framework tower (5 metres)
Established: 1980
Admiralty number: A3996

Light characteristic: Fl W 3s
Grid reference: NF 955 691
Access inside tower: No

Access to site: Very challenging access by land across 3.5kms (2 miles) of extremely difficult terrain from Lochportain. Best viewed by boat. The lighthouse is visible from the Uig ferry on the approach to/departure from Lochmaddy.

4.18 MONACH ISLES NEW (i)

◯ Visited on ....

Next to brick lighthouse on Shillay, Monach Isles

White metal framework tower
Established: 1997
Light discontinued 2008
Grid reference: NF 593 627
Access inside tower: No

Access to site: Access to the island is by chartered boat or organised trips from North Uist. Landing requires great care and calm sea conditions. The operational lighthouse (see 4.19) is very close.

4.19 MONACH ISLES (SHILLAY, SIOLAIGH) (i) ☆

◯ Visited on ....

On Shillay island, Monach Isles

Red brick tower (41 metres)
Established: 1864
Designed by: David and Thomas Stevenson
Admiralty number: A4020.5
Light characteristic: Fl(2)W 15s
Grid reference: NF 593 627
Access inside tower: No

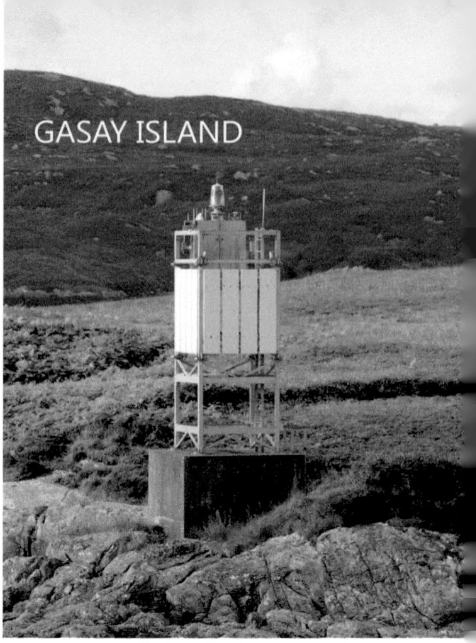

Access to site: Access to the island is by chartered boat or organised trips from North Uist. Landing requires great care and calm sea conditions. The modern light (see 4.18) is very close.

Note: There are plans for the remains of the old lens from the lighthouse to be displayed in the Museum of Scottish Lighthouses, Fraserburgh.

4.20 USHENISH (i) ✯

○ Visited on ....

Eastern point of South Uist

White tower (12 metres)
Established: 1857
Designed by: David and Thomas Stevenson
Admiralty number: A4004
Light characteristic: Fl WR 20s
Grid reference: NF 873 351
Access inside tower: No

Access to site: Accessible by chartered boat only with landing in calm conditions to the east of the lighthouse and a 2km (1.25 mile) walk along a track. Without a boat, access would be possible on foot from Loch Sgioport, involving a 7.5km (4.5 mile) walk on rough ground. The light can also be viewed from the sea by chartered boat or from a distance on a clear day from boats passing through the Sea of the Hebrides.

4.21 GASAY ISLAND (GAASAY, GASAIGH) (i)

○ Visited on ....

South east of Lochboisdale, South Uist

White clad metal framework tower (6 metres)
Admiralty number: A4008.5
Light characteristic: Fl WR 5s

Grid reference: NF 802 189
Access inside tower: No

Access to site: The light is accessible on foot across the causeway from Lochboisdale to Gasay island, followed by a walk across rough terrain. The walk is around 1.5kms (1 mile) from Lochboisdale to the lighthouse. Alternatively, the light can be seen from the Oban/Castlebay to Lochboisdale ferries.

4.22 CALVAY (i)

East coast of Calvay island

○ Visited on ....

White clad metal framework tower (6 metres)
Admiralty number: A4008
Light characteristic: Fl(2)WRG 10s
Grid reference: NF 822 180
Access inside tower: No

Access to site: The light is clearly visible from the Oban/Castlebay to Lochboisdale ferries.

4.23 BARRA HEAD (i) ✯

West coast of Berneray island

○ Visited on ....

White stone tower (18 metres)
Established: 1833
Designed by: Robert Stevenson
Admiralty number: A4020
Light characteristic: Fl W 15s
Grid reference: NL 549 803
Access inside tower: No

Access to site: Access by chartered boat only. Landing requires great care and very calm conditions. From the landing point it is a 2km (1.25 mile) slight uphill walk on a clear track.

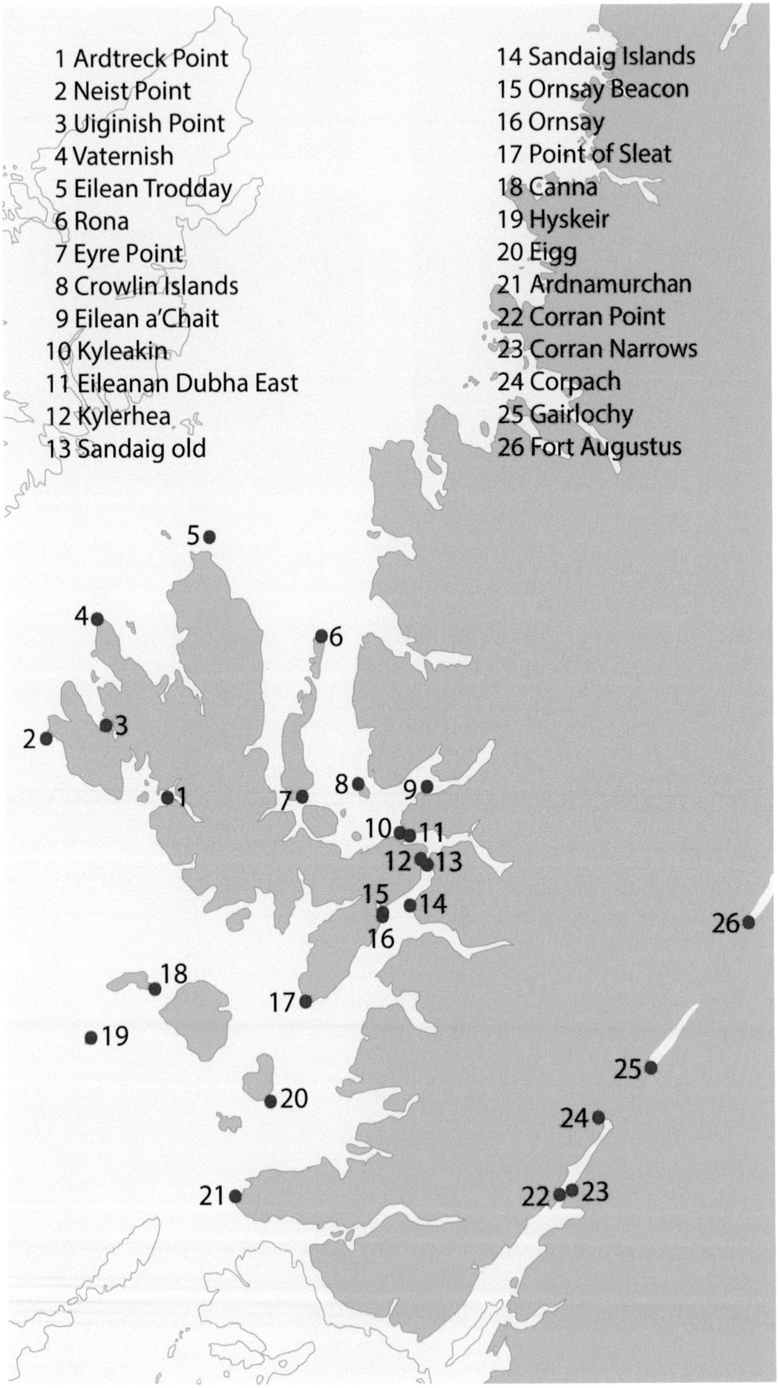
1 Ardtreck Point
2 Neist Point
3 Uiginish Point
4 Vaternish
5 Eilean Trodday
6 Rona
7 Eyre Point
8 Crowlin Islands
9 Eilean a'Chait
10 Kyleakin
11 Eileanan Dubha East
12 Kylerhea
13 Sandaig old
14 Sandaig Islands
15 Ornsay Beacon
16 Ornsay
17 Point of Sleat
18 Canna
19 Hyskeir
20 Eigg
21 Ardnamurchan
22 Corran Point
23 Corran Narrows
24 Corpach
25 Gairlochy
26 Fort Augustus

Region 5

Inner Hebrides and West Highlands

5.1 ARDTRECK POINT (m)

South of entrance to Loch Harport, Skye

White clad metal framework tower (15 metres)
Established: 2002
Admiralty number: A4068
Light characteristic: Fl W 6s
Grid reference: NG 335 364
Access inside tower: No

○ Visited on ....

Access to site: Access is on foot from the end of the road in Ardtreck. The often pathless 1km (0.6 mile) route is over uneven and boggy ground.

5.2 NEIST POINT (m) ☆

Western point of Skye

◯ Visited on ....

White tower (19 metres)
Established: 1909
Designed by: David A and Charles Stevenson
Admiralty number: A4064
Light characteristic: Fl W 5s
Grid reference: NG 126 472
Access inside tower: No

Access to site: Parking is available at the end of a west-bound single-track road off of the B884 near Milovaig. From the car park, it is a 1km (0.6 mile) walk along a clear track to reach the lighthouse. The first section of the walk involves a steep flight of steps.

Note: The old lens from the lighthouse is now on display at the Museum of Scottish Lighthouses, Fraserburgh.

5.3 UIGINISH POINT (DUNVEGAN) (m)

North west of Dunvegan, Skye

◯ Visited on ....

White clad metal framework tower (5 metres)
Admiralty number: A4058
Light characteristic: Fl WRG 3s
Grid reference: NG 235 490
Access inside tower: No

Access to site: From the B884 between Lonmore and Skindin, take the single-track road heading north towards Uiginish. At the end of the public road at Uiginish Farm, continue the 1km (0.6 mile) journey on foot. The track ends a short distance into the walk and the rest is pathless.

5.4 VATERNISH (WATERNISH POINT) (m) ✯

Northern point of Waternish peninsula, Skye

○ Visited on ....

White tower (7 metres)
Established: 1980
Admiralty number: A4056
Light characteristic: Fl W 20s
Grid reference: NG 233 670
Access inside tower: No

Access to site: The lighthouse is accessible on foot from the car park to the north of Trumpan. The 7km (4.5 mile) walk to the lighthouse initially follows the road and then a track. At times the track can be boggy and the final stretch has no clear path.

5.5 EILEAN TRODDAY (i)

Off of northern point of Skye

○ Visited on ....

White clad metal framework tower (7 metres)
Admiralty number: A4050
Light characteristic: Fl(2)WRG 10s
Grid reference: NG 441 790
Access inside tower: No

Access to site: Best seen by chartered boat. The light can be seen from a distance from Skye by following the single-track road, signposted Aird, off of the A855 at Kilmaluag. From the end of the public road it is a 1km (0.6 mile) walk to the top of the hill, which is a good vantage point for seeing the lighthouse.

VATERNISH

EILEAN TRODDAY

5.6 RONA (SOUTH RONA) (i) ☆

○ Visited on

....

North east coast of South Rona

White tower (13 metres)
Established: 1857
Designed by: David and Thomas Stevenson
Admiralty number: A3904
Grid reference: NG 634 612
Light characteristic: Fl W 12s
Access inside tower: No

Access to site: Best seen by chartered boat from Portree on Skye or Raasay. Landing on the island requires prior permission as it is part of a military site. From the landing point it is a short walk on a track towards the north of the island and the lighthouse. The lighthouse can also be seen from the mainland at a distance. The best vantage point is the viewpoint north of Kalnakill on the Applecross to Shieldaig road.

5.7 EYRE POINT (RAASAY) (i)

○ Visited on

....

South eastern point of Raasay

White clad metal framework tower (5 metres)
Established: 2001
Admiralty number: A3909.6
Light characteristic: Fl WR 3s
Grid reference: NG 581 342
Access inside tower: No

Access to site: Scheduled ferries run to Raasay from Sconser on Skye. It is a 5.5km (3.5 mile) drive/walk from the ferry terminal, following the coastal road to the south and east to reach the lighthouse. The lighthouse is a short distance beyond the end of the public road in Eyre.

5.8 CROWLIN ISLANDS (i)

○ Visited on

....

Northern point of Eilean Beag, Crowlin Islands

White clad metal framework tower (5 metres)
Admiralty number: A3916

Light characteristic: Fl W 6s
Grid reference: NG 681 358
Access inside tower: No

Access to site: Best seen by chartered boat. Distance views can be achieved on a clear day from above the village of Toscaig on the mainland or from the south east coast of Raasay.

5.9 EILEAN A'CHAIT (t)

North of Plockton ☆

White square tower (9 metres)
Established: approx. 1880
Light discontinued 1920s
Grid reference: NG 807 347
Access inside tower: No

Access to site: The island is now privately owned but can be accessed by wading at low tide from the headland north of Plockton. Wildlife boat trips from Plockton also give good views of the lighthouse. A distant view of the lighthouse can be achieved from the viewpoint off of Frithard Road to the north west of Plockton.

5.10 KYLEAKIN (EILEAN BÀN) (m) ☆

○ Visited on ....

Under the Skye bridge

White round tower (21 metres)
Established: 1857
Designed by: David and Thomas Stevenson
Light discontinued 1993
Grid reference: NG 745 270

Access inside tower: A visit inside the tower is part of a guided tour, operated by The Bright Water Centre during the summer months. Tickets can be purchased from the Centre in Kyleakin

Access to site: The site is only accessible as part of a tour or during a stay at the cottage. Boat trips operating out of Kyleakin pass the lighthouse and it is also possible to see the lighthouse from the bridge to Skye or from Kyleside in Kyleakin village.

Note: The old lens from the lighthouse is now on display in the Museum of Scottish Lighthouses, Fraserburgh.

5.11 EILEANAN DUBHA EAST (i)

○ Visited on ....

Off of coast at Kyle of Lochalsh

White clad metal framework tower (5 metres)
Admiralty number: A3927
Light characteristic: Fl(2)W 10s
Grid reference: NG 767 267
Access inside tower: No

Access to site: The light can be seen from the viewpoint to the south of Kyle of Lochalsh on the A87. A closer view would require a chartered boat.

5.12 KYLERHEA (t) ✯

○ Visited on ....

North of Kylerhea ferry terminal, Skye

White tower (7 metres)
Established: 1892
Designed by: David A and Charles Stevenson
Admiralty number: A3938
Light characteristic: Fl WRG 3s
Grid reference: NG 789 223
Access inside tower: No

Access to site: Public parking available at the Kylerhea Otter Haven. From here it is a 1.5km (1 mile) walk along tarmac road and track to reach the lighthouse. The old Sandaig lighthouse (see 5.13) is visible from here.

5.13 SANDAIG OLD (m) ✯

○ Visited on ....

Relocated to slipway at Glenelg ferry from Sandaig Islands

White octagonal cast iron tower (8 metres)
Established: 1910
Designed by: David A and Charles Stevenson

KYLERHEA

Light discontinued 2004
Grid reference: NG 795 213
Access inside tower: The lighthouse now contains tourist information and details of the Glenelg-Kylerhea ferry.

Access to site: Accessible by road along 3km (1.8 mile) single-track road from Glenelg. Also accessible by taking the ferry from Kylerhea on Skye. The Kylerhea light (see 5.12) is visible from the lighthouse.

5.14 SANDAIG ISLANDS (i)

○ Visited on ....

On Eilean Mòr, Sandaig islands

White clad metal framework tower (5 metres)
Established: 2004
Admiralty number: A3940
Light characteristic: Fl W 6s
Grid reference: NG 762 146
Access inside tower: No

Access to site: Best seen by chartered boat. It is possible to see the lighthouse from the tidal islands within the Sandaig island group. This can be achieved by parking at the roadside on the main road through Sandaig and following the track down to the coast.

Note: For details of the former lighthouse in this location see 5.13.

5.15 ORNSAY BEACON (t)

○ Visited on ....

Off of north end of Ornsay island, east coast of Skye

White clad metal framework tower on stone column (8 metres)
Established: 2002
Admiralty number: A3942
Light characteristic: Fl R 6s
Grid reference: NG 714 130
Access inside tower: No

Access to site: Ornsay is accessible at low tide across the beach from Isleornsay. The crossing can be wet underfoot. Once on the island it is a short walk to the north and east to reach the light. The light itself may be accessible with care at the lowest of tides.

SANDAIG OLD

ORNSAY BEACON

5.16 ORNSAY (EILEAN SIONNACH) (t) ✯

South of Ornsay, east coast of Skye

○ Visited on ....

White tower (19 metres)
Established: 1857
Designed by: David and Thomas Stevenson
Admiralty number: A3944
Light characteristic: Oc W 8s
Grid reference: NG 714 122
Access inside tower: No

Access to site: Ornsay is accessible at low tide across the beach from Isleornsay. The crossing can be wet underfoot, followed by a walk eastwards across rocks on the main island. Once on Eilean Sionnach it is a short, easy walk to the lighthouse.

5.17 POINT OF SLEAT (m)

Southern point of Skye

○ Visited on ....

White clad metal framework tower (7 metres)
Established: 2003
Admiralty number: A3952
Light characteristic: Fl W 3s
Grid reference: NM 562 991
Access inside tower: No

Access to site: Public parking is available at the end of the public road at Armadale. A clearly marked track then takes you on the 4.5km (3 mile) walk to the lighthouse. Some sections of the walk are rougher underfoot.

5.18 CANNA (SANDAY) (i) ☆

East coast of Sanday island

○ Visited on ....

White tower (9 metres)
Admiralty number: A4074
Light characteristic: Fl W 10s
Grid reference: NG 292 040
Access inside tower: No

Access to site: Scheduled ferries run from Mallaig to Canna. Once on Canna it is a 5km (3 mile) walk, via the bridge across to Sanday, to reach the lighthouse. Beyond St Edwards Centre on Sanday the terrain becomes rougher. The lighthouse can also be seen on a clear day from the ferry.

5.19 HYSKEIR (OIGH SGEIR) (i) ☆

On Oigh Sgeir island, west of Rum

○ Visited on ....

White tower (39 metres)
Established: 1904
Designed by: David A and Charles Stevenson
Admiralty number: A4076
Light characteristics: Fl(3)W 30s
Grid reference: NM 156 962
Access inside tower: No

Access to site: A close view of the lighthouse requires a chartered boat. Landing is possible in calm sea conditions. The lighthouse can be seen from a distance from the Oban to Barra ferry.

EIGG

5.20 EIGG (EILEAN CHATHASTAIL) (i) ☆

South east coast of Eilean Chathastail, south east of Eigg

○ Visited on ....

White tower (8 metres)
Light characteristic: Fl W 6s
Grid reference: NM 489 831
Access inside tower: No

Access to site: Landing on the island is possible from a chartered boat. Landing is only possible in calm sea conditions and requires a short scramble on rocks to reach the lighthouse. The light is visible from scheduled ferries between Eigg and Muck.

5.21 ARDNAMURCHAN (m) ☆

At end of Ardnamurchan peninsula

○ Visited on ....

Grey granite tower (36 metres)
Established: 1849
Designed by: Alan Stevenson
Admiralty number: A4082
Light characteristic: Fl(2)W 20s
Grid reference: NM 416 675
Access to tower: Tours of the tower run daily during the summer months.

Access to site: The lighthouse is accessible by road, 56kms (35 miles) west of Strontian. There is parking available at the lighthouse. The lighthouse is visible from ferries between Oban and Castlebay or Lochboisdale.

Note: The former lens from the lighthouse is now on display in the Ardnamurchan Lighthouse Visitor Centre.

ARDNAMURCHAN

5.22 CORRAN POINT (m) ☆

○ Visited on ....

Near ferry terminal in Ardgour

White tower (13 metres)
Established: 1860
Designed by: David and Thomas Stevenson
Admiralty number: A4132
Light characteristic: Iso WRG 4s
Grid reference: NN 017 635
Access inside tower: No

Access to site: The site is privately owned, but the lighthouse is visible from the A861 as well as the beach in front of the lighthouse and the Corran ferry.

5.23 CORRAN NARROWS (m)

○ Visited on ....

North of Corran

White metal framework tower (4 metres)
Admiralty number: A4133
Light characteristic: Fl W 5s
Grid reference: NN 024 641
Access inside tower: No

Access to site: Access is via a short walk downhill through trees from the northbound sharp corner sign on the A82 just north of Corran. Parking is available at the side of the A82 just to the north of the track.

5.24 CORPACH (m)

○ Visited on ....

Caledonian Canal lock entrance

White tower, black top (6 metres)
Established: 1913
Admiralty number: A4140

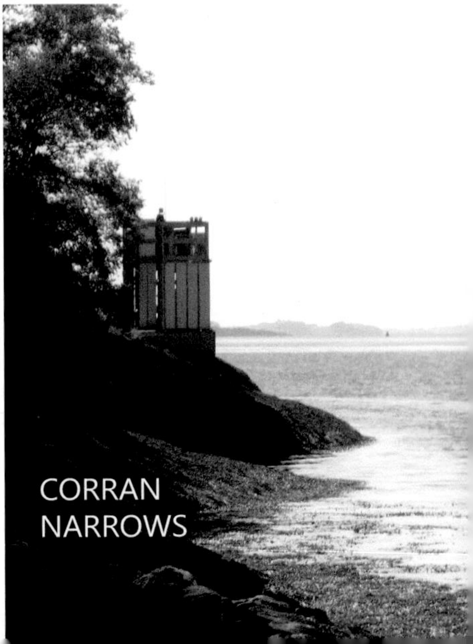

Light characteristic: Iso WRG 4s
Grid reference: NN 096 767
Access inside tower: No

Access to site: Public parking is available on the north bank of the Caledonian Canal at Corpach. From here a short walk along the road heading west takes you to the lighthouse.

5.25 GAIRLOCHY (m)

Northern entrance to Caledonian Canal, north of Gairlochy

○ Visited on ....

White tower (6 metres)
Established: 1932
Admiralty number: A3480
Light characteristic: Fl WG 3s
Grid reference: NN 183 846
Access inside tower: No

Access to site: Access on foot along the towpath from the B8004 at Gairlochy, following the east bank of the Caledonian Canal. There is a small parking area opposite the entrance to the towpath.

5.26 FORT AUGUSTUS (m)

South east bank of Fort Augustus, southern end of Loch Ness

○ Visited on ....

White tower (6 metres)
Admiralty number: A3472
Light characteristic: Fl G 3s
Grid reference: NH 383 093
Access inside tower: No

Access to site: Public parking off of A82 north of Caledonian Canal crossings at Fort Augustus. From here, walk across the bridges and then along the south bank of the canal to reach the lighthouse.

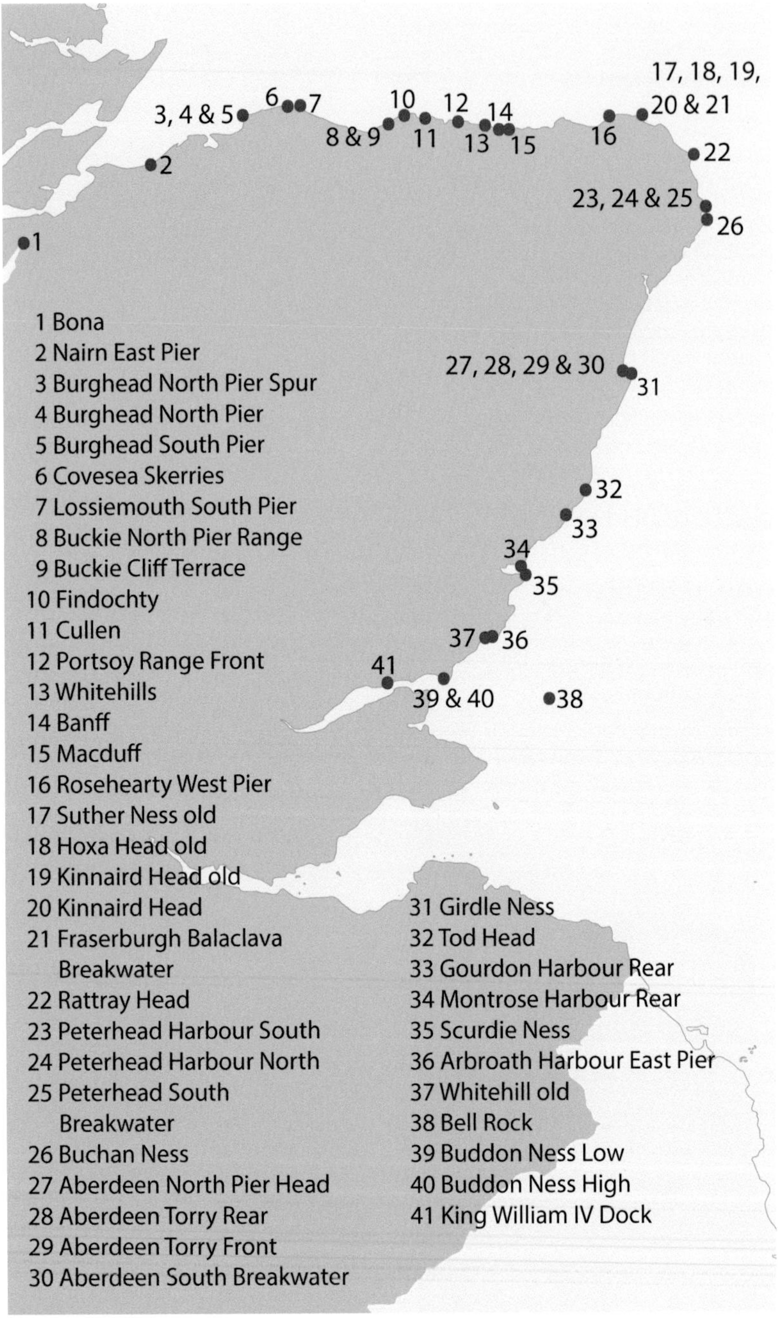

17, 18, 19, 20 & 21
3, 4 & 5
6
7
8 & 9
10
11
12
13
14
15
16
2
22
23, 24 & 25
26
1
27, 28, 29 & 30
31
32
33
34
35
37
36
41
39 & 40
38
1 Bona
2 Nairn East Pier
3 Burghead North Pier Spur
4 Burghead North Pier
5 Burghead South Pier
6 Covesea Skerries
7 Lossiemouth South Pier
8 Buckie North Pier Range
9 Buckie Cliff Terrace
10 Findochty
11 Cullen
12 Portsoy Range Front
13 Whitehills
14 Banff
15 Macduff
16 Rosehearty West Pier
17 Suther Ness old
18 Hoxa Head old
19 Kinnaird Head old
20 Kinnaird Head
21 Fraserburgh Balaclava Breakwater
22 Rattray Head
23 Peterhead Harbour South
24 Peterhead Harbour North
25 Peterhead South Breakwater
26 Buchan Ness
27 Aberdeen North Pier Head
28 Aberdeen Torry Rear
29 Aberdeen Torry Front
30 Aberdeen South Breakwater
31 Girdle Ness
32 Tod Head
33 Gourdon Harbour Rear
34 Montrose Harbour Rear
35 Scurdie Ness
36 Arbroath Harbour East Pier
37 Whitehill old
38 Bell Rock
39 Buddon Ness Low
40 Buddon Ness High
41 King William IV Dock

Region 6

North East Scotland

6.1 BONA (m)

North end of Loch Ness

○ Visited on

....

White octagonal tower with adjoining single storey building (10 metres)
Light discontinued
Grid reference: NH 602 377
Access inside tower: Access is for those staying at the accommodation only.

Access to site: The tower is accessible from the turn off of the A82 at Lochend. Where the public road ends, unless staying at the lighthouse accommodation, a short walk along the road takes you to the lighthouse. The building is also visible from boat trips into Loch Ness.

6.2 NAIRN EAST PIER (m)

In Nairn harbour

Octagonal concrete tower (5 metres)
Established: 1932
Admiralty number: A3436
Light characteristic: Oc WRG 4s
Grid reference: NH 889 575
Access inside tower: No

○ Visited on

....

Access to site: Public parking is available on Maggot Road, Nairn. From the car park a short walk north along the east bank of the River Nairn takes you to the pier.

6.3 BURGHEAD NORTH PIER SPUR (m)

○ Visited on ....

At end of Burghead north pier

Concrete structure (3 metres)
Admiralty number: A3426
Light characteristic: Q R
Grid reference: NJ 107 689
Access inside tower: No

Access to site: Parking is available on Station Road. A short walk from here takes you to the harbour. You pass the north pier light (see 6.4) to reach the north pier spur. The south pier light (see 6.5) is also visible from here.

6.4 BURGHEAD NORTH PIER (m)

○ Visited on ....

In Burghead harbour

Concrete tower (6 metres)
Established: 1883
Admiralty number: A3424
Light characteristic: Oc W 8s
Grid reference: NJ 107 690
Access inside tower: No

Access to site: Parking is available on Station Road. A short walk from here takes you to the harbour. The north pier spur light (see 6.3) is a little further on from here and the south pier light (see 6.5) is also visible from here.

6.5 BURGHEAD SOUTH PIER (m)

○ Visited on ....

At end of Burghead south pier

Wooden structure (3 metres)
Admiralty number: A3428
Light characteristic: Q G
Grid reference: NJ 108 690
Access inside tower: No

Access to site: Parking is available on Station Road. A short walk from here, via Church Street, takes you to the south pier. The north pier (see 6.4) and north pier spur (see 6.3) lights are visible from here.

6.6 COVESEA SKERRIES (m) ✯

○ Visited on ....

West of Lossiemouth

White tower (36 metres)
Established: 1846
Designed by: Alan Stevenson
Light discontinued 2012
Grid reference: NJ 204 713

Access inside tower: Access is possible by guided tour. The tours run on Saturday mornings from April to October, although arrangements can be made outside of these times by appointment. It is advisable to book in advance and more details, as well as contact details, can be found at www.covesealighthouse.co.uk

Access to site: Access by road is just inside the entrance to Silver Sands holiday park from the B9040. Parking is available near the lighthouse.

Note: The old lens from the lighthouse is now on display at the nearby Lossiemouth Fisheries and Community Museum.

6.7 LOSSIEMOUTH SOUTH PIER (m)

○ Visited on ....

In Lossiemouth harbour

Square skeletal tower with gallery, enclosed lower section (11 metres)

COVESEA SKERRIES

LOSSIEMOUTH SOUTH PIER

Admiralty number: A3408
Light characteristic: Fl R 6s
Grid reference: NJ 240 711
Access inside tower: No

Access to site: Public parking is available on Pitgaveny Street. The light is visible from here and can be accessed by a short walk around the harbour.

6.8 BUCKIE NORTH PIER RANGE (m) ✯

On north pier at Buckie

○ Visited on ....

White tower (15 metres)
Established: 1913
Admiralty number: A3394
Light characteristic: Oc R 10s
Grid reference: NJ 429 661
Access inside tower: No

Access to site: The pier is accessible at the end of Blantyre Terrace.

6.9 BUCKIE CLIFF TERRACE (m) ✯

On Cliff Terrace, Buckie

○ Visited on ....

White column (10 metres)
Established: 1878
Admiralty number: A3396.1
Light characteristic: Iso WG 2s
Grid reference: NJ 431 658
Access inside tower: No

Access to site: The lighthouse is at the side of the road. Parking is available nearby.

6.10 FINDOCHTY (m)

At end of west pier in Findochty harbour

○ Visited on ....

Round concrete tower, gold cast iron lantern (10 metres)
Established: 1904
Light discontinued
Grid reference: NJ 462 680
Access inside tower: No

Access to site: The west pier is accessible on foot. Public parking is available on Jubilee Terrace near the entrance to the pier.

6.11 CULLEN (m)

At end of east pier, Cullen

○ Visited on ....

White round tower, domed roof (10 metres)
Established: 1866
Admiralty number: A3372
Light characteristic: F G
Grid reference: NJ 510 675
Access inside tower: No

Access to site: The east pier is accessible on foot. It is a short walk to the east pier from the beach road where parking is available.

6.12 PORTSOY RANGE FRONT (m)

In Portsoy harbour

○ Visited on ....

White round tower, cast iron lantern (8 metres)
Established: approx. 1884
Light discontinued 2009

Grid reference: NJ 590 663
Access inside tower: No

Access to site: The lighthouse is on Shore Street with parking a short distance away.

6.13 WHITEHILLS (m)

○ Visited on ....

At end of east pier, Whitehills

White tower (3 metres)
Established: 1900
Admiralty number: A3364
Light characteristic: Fl WR 3s
Grid reference: NJ 655 657
Access inside tower: No

Access to site: Access to the east pier is next to Blackpots Recreation Ground. There is parking nearby.

6.14 BANFF (m)

○ Visited on ....

On north pier, Banff

White round lantern atop a stone pier head building (8 metres)
Established: 1832
Admiralty number: A3353
Light characteristic: Fl W 4s
Grid reference: NJ 689 647
Access inside tower: No

Access to site: The pier is accessible on foot. Public parking is available on Quayside, a short walk away.

6.15 MACDUFF (m) ✯

○ Visited on ....

At pier head, Macduff

White tower (11 metres)
Established: 1905
Admiralty number: A3348
Light characteristic: Fl(2)WRG 6s
Grid reference: NJ 703 647
Access inside tower: No

Access to site: Access to the pier is at the end of Laing Street, beyond the lifeboat station. Parking is available on Low Shore.

6.16 ROSEHEARTY WEST PIER (m)

○ Visited on ....

At end of west pier, Rosehearty

White round tower (5 metres)
Established: 1883
Admiralty number: A3335
Light characteristic: Fl W 4s
Grid reference: NJ 932 679
Access inside tower: No

Access to site: Access to the pier is from Union Street.

6.17 SUTHER NESS OLD (m) ✯

○ Visited on ....

Relocated from Suther Ness, Shetland, to Museum of Scottish Lighthouses, Fraserburgh

White round tower
Established: 1904

Designed by: David A and Charles Stevenson
Light discontinued
Grid reference: NJ 997 675
Access inside tower: No

Access to site: The lighthouse is in the car park at the Museum of Scottish Lighthouses. A visit to the museum also takes in the Hoxa Head old light (see 6.18), the old Kinnaird Head lighthouse (see 6.19) and the current Kinnaird Head light (see 6.20).

6.18 HOXA HEAD OLD (m) ☆

○ Visited on ....

Relocated from Hoxa Head, Orkney, to Museum of Scottish Lighthouses, Fraserburgh

White round tower
Established: 1901
Designed by: David A and Charles Stevenson
Light discontinued
Grid reference: NJ 998 676
Access inside tower: Yes. The tower contains part of the lighthouse exhibition.

Access to site: The tower is inside the exhibition room at the museum which is open daily throughout the summer months and 6 days per week in winter. A visit to the museum also takes in the old Suther Ness light (see 6.17), the old Kinnaird Head lighthouse (see 6.19) and the current Kinnaird Head light (see 6.20).

6.19 KINNAIRD HEAD OLD (m) ☆

○ Visited on ....

In grounds of Museum of Scottish Lighthouses, Fraserburgh

White tower atop 16th century stone castle (22 metres)
Established: 1787
Designed by: Thomas Smith
Light discontinued 1991
Grid reference: NJ 999 675
Access inside tower: Yes. There are daily tours of the tower.

Access to site: The lighthouse is within the grounds of the Museum of Scottish Lighthouses. A visit to the museum also takes in the old Suther Ness light (see 6.17), the old Hoxa Head lighthouse (see 6.18) and the new Kinnaird Head light (see 6.20). It is worth allowing plenty of time to visit Kinnaird Head lighthouse and the Museum.

6.20 KINNAIRD HEAD (m) ☆

○ Visited on ....

North end of Fraserburgh, in grounds of Museum of Scottish Lighthouses

White tower (10 metres)
Established: 1991
Admiralty number: A3332
Light characteristic: Fl W 5s
Grid reference: NJ 999 676
Access inside tower: No

Access to site: Within the grounds of the Museum of Scottish Lighthouses. A visit to the museum also takes in the old Suther Ness light (see 6.17), the old Hoxa

Head lighthouse (see 6.18) and the old Kinnaird Head light (see 6.19). It is worth allowing plenty of time to visit Kinnaird Head lighthouse and the Museum.

6.21 FRASERBURGH BALACLAVA BREAKWATER (m) ☆

○ Visited on ....

At end of north breakwater, Fraserburgh

Dome on white stone tower, red band (21 metres)
Established: 1882
Admiralty number: A3312
Light characteristic: Fl(2)G 8s
Grid reference: NK 004 669
Access inside tower: No

Access to site: The breakwater is accessible on foot from the Museum of Scottish Lighthouses. A short walk takes you to and along the breakwater. There is no public access to the end of the breakwater and the lighthouse, but it can be seen from the gates.

6.22 RATTRAY HEAD (t) ☆

North of Peterhead

○ Visited on ....

White tower, lower part granite, upper brick (34 metres)
Established: 1895
Designed by: David A and Charles Stevenson
Admiralty number: A3304
Light characteristic: Fl(3)W 30s
Grid reference: NK 111 579
Access inside tower: No

Access to site: The tower is accessible at low tide. Public parking is available at the end of the single-track road to the north west of Rattray. From the car park it is a short walk over the dunes to the beach. The access road may be in poor condition.

Note: The old lens from the lighthouse can now be seen at Aberdeen Maritime Museum.

6.23 PETERHEAD HARBOUR SOUTH (m) ☆

Relocated from Peterhead harbour to Alexandra Parade, Peterhead

○ Visited on ....

Octagonal stone tower, white lantern (5 metres)
Established: 1849
Light discontinued
Grid reference: NK 136 466
Access inside tower: No

Access to site: The lighthouse is at the side of Alexandra Parade, at the junction with The Esplanade.

6.24 PETERHEAD HARBOUR NORTH (m) ☆

Next to Peterhead Port Authority building

○ Visited on ....

Octagonal stone tower, white lantern (5 metres)
Established: 1908
Light discontinued

Grid reference: NK 136 458
Access inside tower: No

Access to site: Parking is available on the West Pier next to Port Authority building. A walk around the outside of the Port Authority building takes you to the lighthouse.

6.25 PETERHEAD SOUTH BREAKWATER (m)

South of entrance to Peterhead harbour

○ Visited on ....

White round metal tower, black base (17 metres)
Established: 1833
Admiralty number: A3284
Light characteristic: Fl(2)R 12s
Grid reference: NK 135 452
Access inside tower: No

Access to site: Access to the breakwater is not permitted. The lighthouse can be seen from a distance from South Road car park.

6.26 BUCHAN NESS (m) ✯

On headland in Boddam

○ Visited on ....

White tower, red band (35 metres)
Established: 1827
Designed by: Robert Stevenson
Admiralty number: A3280

Light characteristic: Fl W 5s
Grid reference: NK 137 422
Access inside tower: No

Access to site: The lighthouse is accessible on foot from Bridge Street, Boddam across the bridge.

Note: The former lens from the lighthouse is now currently in storage at the Museum of Scottish Lighthouses, Fraserburgh.

6.27 ABERDEEN NORTH PIER HEAD (m) ☆

○ Visited on ....

At end of north pier, Aberdeen

White tower (9 metres)
Established: 1866
Admiralty number: A3252
Light characteristic: Iso G 4s
Grid reference: NJ 964 061
Access inside tower: No

Access to site: Access to the pier is not permitted. The lighthouse is best seen from Greyhope Road to the south of the harbour entrance, close to the south breakwater light (see 6.30) and Girdle Ness (see 6.31). The light can also be seen from ferries to or from Kirkwall and Lerwick.

6.28 ABERDEEN TORRY REAR (m) ☆

○ Visited on ....

South side of Aberdeen harbour

White tower (14 metres)
Established: 1842
Admiralty number: A3258.1
Light characteristic: F R
Grid reference: NJ 954 053
Access inside tower: No

Access to site: The lighthouse is at the side of Sinclair Road. The front light (see 6.29) is a short distance away.

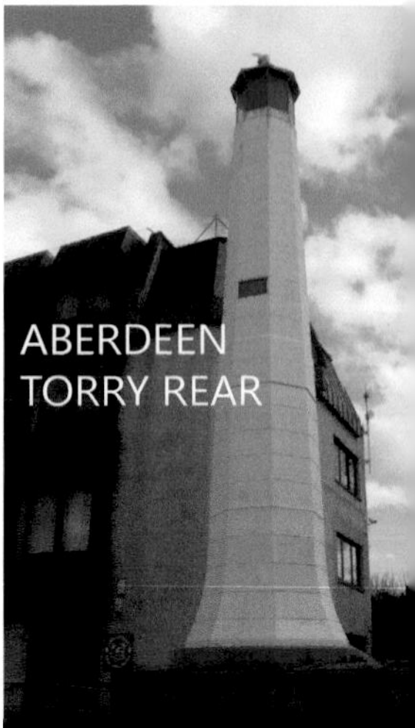

6.29 ABERDEEN TORRY FRONT (m) ☆

South side of Aberdeen harbour, east of Torry Rear light

○ Visited on

....

White tower (13 metres)
Established: 1842
Admiralty number: A3258
Light characteristic: F R
Grid reference: NJ 956 054
Access inside tower: No

Access to site: The lighthouse is at the side of Sinclair Road. The rear light (see 6.28) is a short distance away.

6.30 ABERDEEN SOUTH BREAKWATER HEAD (m)

At end of south breakwater, Aberdeen

○ Visited on

....

White tower (approx. 13 metres)
Established: 1866
Admiralty number: A3250
Light characteristic: Fl(3)R 8s
Grid reference: NJ 967 060
Access inside tower: No

Access to site: Access to the breakwater is restricted. Good views of the lighthouse are possible from Greyhope Road and the surrounding area. The north pier light (see 6.27) and Girdle Ness (see 6.31) can also be seen from here. The light can also be seen from ferries to or from Kirkwall and Lerwick.

6.31 GIRDLE NESS (m) ☆

South of Aberdeen harbour

○ Visited on

....

White round tower (37 metres)
Established: 1833
Designed by: Robert Stevenson
Admiralty number: A3246
Light characteristic: Fl(2)W 20s

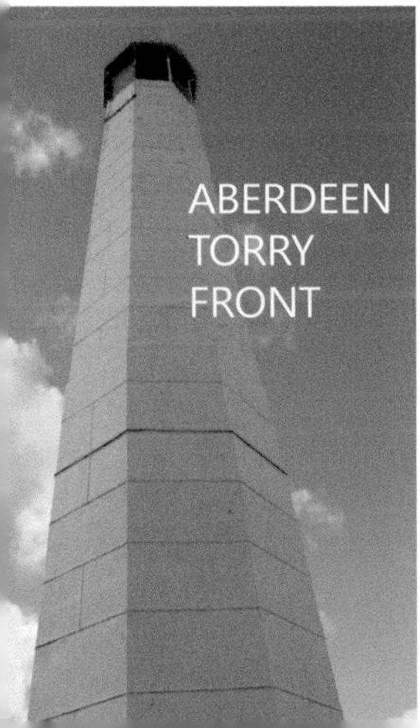

Grid reference: NJ 971 053
Access inside tower: No

Access to site: Clearly visible from Greyhope Road where public parking is available. The north pier (see 6.27) and south breakwater (see 6.30) lights can both be seen by taking a short walk north west from the lighthouse.

6.32 TOD HEAD (m) ☆

South of Catterline

○ Visited on ....

White tower (13 metres)
Established: 1897
Designed by: David A and Charles Stevenson
Light discontinued 2007
Grid reference: NO 870 770
Access inside tower: Visiting the tower is possible on special occasions only.

Access to site: The site is accessible from a turn-off of the single-track road between Catterline and Kinneff.

6.33 GOURDON HARBOUR REAR (m)

On Brae Road, Gourdon

○ Visited on ....

White tower (6 metres)
Established: 1938
Admiralty number: A3228.1
Light characteristic: F R
Grid reference: NO 826 709
Access inside tower: No

Access to site: The lighthouse is visible and easily accessible from Brae Road.

6.34 MONTROSE HARBOUR REAR (m) ☆

North bank of River South Esk, Montrose

○ Visited on ....

White tower, red cupola (19 metres)
Established: 1818
Admiralty number: A3222.1
Light characteristic: F R
Grid reference: NO 719 570
Access inside tower: No

Access to site: The light is accessible down a lane next to Jack Smith Road, off of Cobden Street.

6.35 SCURDIE NESS (m) ✯

South of entrance to River South Esk

○ Visited on ....

White tower (39 metres)
Established: 1870
Designed by: David and Thomas Stevenson
Admiralty number: A3220
Light characteristic: Fl(3)W 20s
Grid reference: NO 734 568
Access to tower: No

Access to site: Parking is available at the end of River Street, Ferryden, followed by a 2km (1.25 mile) walk eastwards along a path and tarmac road.

6.36 ARBROATH HARBOUR EAST PIER (m)

North of entrance to Arbroath harbour

○ Visited on ....

White tower (9 metres)
Admiralty number: A3216
Light characteristic: Fl G 3s
Grid reference: NO 643 404
Access inside tower: No

Access to site: The pier is accessible on foot from Old Shore Head.

6.37 WHITEHILL (VATSETTER) OLD (m) ✯

Relocated to side of road to south west of Arbroath from Whitehill, Shetland

○ Visited on ....

Red round cast iron tower (4.5 metres)
Established: 1904
Designed by: David A and Charles Stevenson
Light discontinued 2001
Grid reference: NO 622 397
Access inside tower: No

Access to site: The tower can be seen at the side of the A92 with parking just beyond the lighthouse when heading into Arbroath.

6.38 BELL ROCK (r) ✯

East of Arbroath

◯ Visited on ....

White round tower (36 metres)
Established: 1811
Designed by: Robert Stevenson
Admiralty number: A3108
Light characteristic: Fl W 5s
Grid reference: NO 762 270
Access to tower: No

Access to site: Regular boat trips to the lighthouse operate from Arbroath.

Note: The old signal tower for Bell Rock is in Arbroath and is now open as a museum, including Bell Rock exhibits such as the last remaining panels of the old lens.

6.39 BUDDON NESS LOW (m) ✯

East of Monifieth

◯ Visited on ....

Round brick and stone tower (14 metres)
Established: 1866
Designed by: David and Thomas Stevenson
Light discontinued 1943
Grid reference: NO 542 308
Access inside tower: No

Access to site: The lighthouse is within the Barry Buddon Firing Range. When live firing is taking place there is no access to the lighthouse. When firing is not taking place, from the main entrance at Barry Links station it is a 3km (1.8 mile) walk to the lighthouse, passing the high light (see 6.40).

6.40 BUDDON NESS HIGH (m) ✯

East of Monifieth

◯ Visited on ……………………

White round brick and stone tower (14 metres)
Established: 1866
Designed by: David and Thomas Stevenson
Light discontinued 1943
Grid reference: NO 539 310
Access inside tower: No

Access to site: The lighthouse is within the Barry Buddon Firing Range. When live firing is taking place there is no access to the lighthouse. When firing is not taking place, from the main entrance at Barry Links station it is a 2.8km (1.75 mile) walk to the lighthouse. A further short walk takes you to the low lighthouse (see 6.39).

6.41 KING WILLIAM IV DOCK (TELFORD BEACON) (m) ✯

Relocated from King William IV Dock to west of Tay Road Bridge, Dundee

◯ Visited on ……………………

Round stone tower (7 metres)
Light discontinued 1960s
Grid reference: NO 407 301
Access inside tower: No

Access to site: The lighthouse can be reached by taking the turn off of the South Victoria Dock Road just to the east of the Tay Road Bridge. The lighthouse is clearly visible as you emerge from under the bridge.

KING WILLIAM IV DOCK

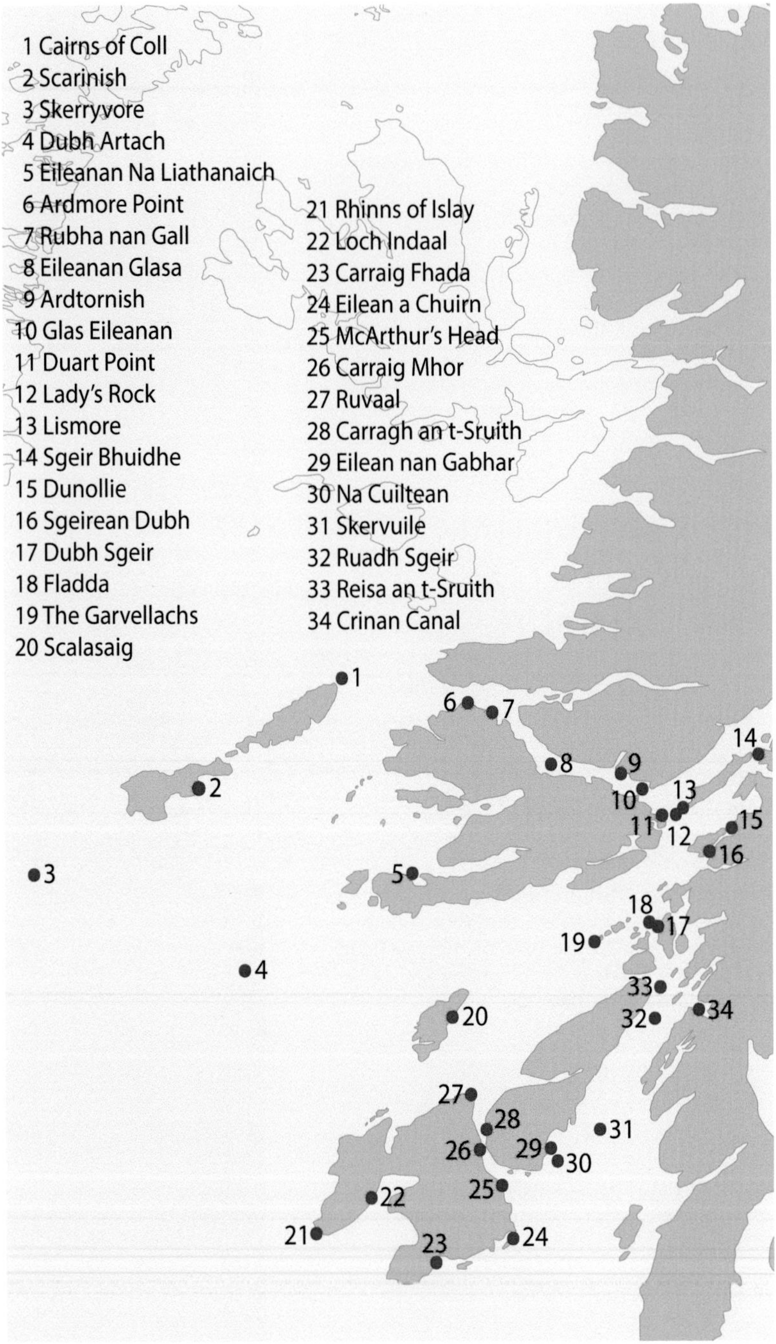
1 Cairns of Coll
2 Scarinish
3 Skerryvore
4 Dubh Artach
5 Eileanan Na Liathanaich
6 Ardmore Point
7 Rubha nan Gall
8 Eileanan Glasa
9 Ardtornish
10 Glas Eileanan
11 Duart Point
12 Lady's Rock
13 Lismore
14 Sgeir Bhuidhe
15 Dunollie
16 Sgeirean Dubh
17 Dubh Sgeir
18 Fladda
19 The Garvellachs
20 Scalasaig
21 Rhinns of Islay
22 Loch Indaal
23 Carraig Fhada
24 Eilean a Chuirn
25 McArthur's Head
26 Carraig Mhor
27 Ruvaal
28 Carragh an t-Sruith
29 Eilean nan Gabhar
30 Na Cuiltean
31 Skervuile
32 Ruadh Sgeir
33 Reisa an t-Sruith
34 Crinan Canal

Region 7

Mull To Islay

7.1 CAIRNS OF COLL (SUIL GHORM) (i) ✯

North east of Coll

White tower (8 metres)
Admiralty number: A4084
Light characteristic: Fl W 12s
Grid reference: NM 280 658
Access inside tower: No

○ Visited on ....

Access to site: Best seen by chartered boat. The lighthouse is visible from a distance from ferries between Oban and Coll or Castlebay.

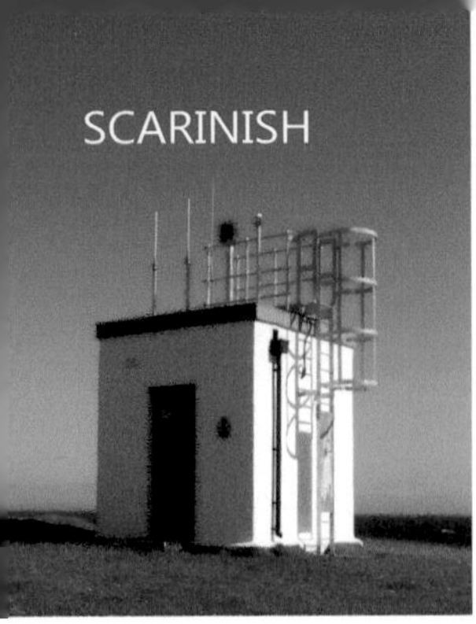

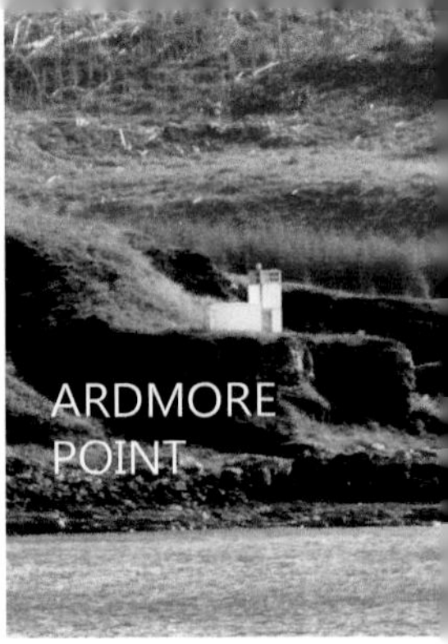

7.2 SCARINISH (i)

South of Gott Bay, Tiree

○ Visited on ....

White square tower (3 metres)
Admiralty number: A4092
Light characteristic: Fl W 3s
Grid reference: NM 045 448
Access inside tower: No

Access to site: Tiree can be reached by scheduled ferry from Oban, or by plane from Glasgow. Once on the island, the light is a short distance from the Scarinish Hotel.

7.3 SKERRYVORE (r) ☆

South west of Tiree

○ Visited on ....

Grey granite tower (48 metres)
Established: 1844
Designed by: Alan Stevenson
Admiralty number: A4096
Light characteristic: Fl W 10s
Grid reference: NL 839 262
Access inside tower: No

Access to site: Regular boat trips run from Tiree during the summer months. The lighthouse can be seen on a clear day from Tiree, or on an exceptionally clear day from the old signal tower on Erraid, which is accessible at low tide from Mull.

Note: The old shore station and signal tower for Skerryvore can be visited at Hynish on the south of Tiree.

7.4 DUBH ARTACH (DHU HEARTACH) (r) ☆

North west of Colonsay

○ Visited on ....

Grey granite tower, red band (38 metres)
Established: 1872
Designed by: David and Thomas Stevenson
Admiralty number: A4098
Light characteristic: Fl(2)W 30s
Grid reference: NM 121 029
Access inside tower: No

Access to site: Best seen by chartered boat. Landing on the rock requires exceptionally calm sea conditions. The lighthouse can be seen on a clear day from the old signal tower on Erraid, which is accessible at low tide from Mull.

7.5 EILEANAN NA LIATHANAICH (BUNESSAN) (i)

○ Visited on ....

North of the Ross of Mull

White clad metal framework tower (5 metres)
Established: 2001
Admiralty number: A4102
Light characteristic: Fl WR 6s
Grid reference: NM 361 249
Access inside tower: No

Access to site: Mull is accessible by vehicle ferry from Craignure, Lochaline or Kilchoan. The lighthouse is visible from the road heading north from Bunessan. A closer view would require a chartered boat.

7.6 ARDMORE POINT (i)

○ Visited on ....

Northern point of Mull

White clad metal framework tower (5 metres)
Established: 2003
Admiralty number: A4107
Light characteristic: Fl(2)W 10s
Grid reference: NM 470 592
Access inside tower: No

Access to site: Parking is available at the Ardmore Forest car park, on the road to the west of Tobermory. From here it is a 4km (2.5 mile) walk following a clear path around the forest and then heading north to reach the lighthouse.

7.7 RUBHA NAN GALL (i) ✯

North of Tobermory, Mull

○ Visited on ....

White tower (19 metres)
Established: 1857
Designed by: David and Thomas Stevenson
Admiralty number: A4112
Light characteristic: Fl W 3s
Grid reference: NM 508 571
Access inside tower: No

Access to site: The lighthouse is accessible on foot via a 2km (1.25 mile) coastal footpath north from the lifeboat station in Tobermory.

7.8 EILEANAN GLASA (GREEN ISLANDS, DEARG SGEIR) (i)

In the Sound of Mull

○ Visited on ....

White clad metal framework tower (7 metres)
Established: 2001
Admiralty number: A4122
Light characteristic: Fl W 6s
Grid reference: NM 595 452
Access inside tower: No

Access to site: Best seen by chartered boat. The lighthouse is visible from ferry sailings from Oban to Coll/Tiree, South Uist and Barra. The light is also visible from a distance from the B849 north west of Fiunary, or from various points along the road around the Salen area on Mull.

7.9 ARDTORNISH (m)

South east of Lochaline

○ Visited on ....

White clad metal framework tower (5 metres)
Admiralty number: A4124

Light characteristic: Fl(2)WRG 10s
Grid reference: NM 692 425
Access inside tower: No

Access to site: The lighthouse is visible from sailings from Oban to Coll/Tiree, South Uist and Barra. Walking to the lighthouse involves a 6.5km (4 mile) walk heading south from Ardtornish along a road most of the way.

7.10 GLAS EILEANAN (GREY ROCKS) (i)

At south eastern entrance to Sound of Mull

○ Visited on ....

White clad metal framework tower (6 metres)
Established: 2001
Admiralty number: A4126
Light characteristic: Fl W 3s
Grid reference: NM 715 400
Access inside tower: No

Access to site: Best seen by chartered boat. The lighthouse is visible from sailings from Oban to Coll/Tiree, South Uist and Barra. The light is also visible from a distance from the A849 on Mull to the north west of Criagnure.

7.11 DUART POINT (i)

Eastern point of Mull

○ Visited on ....

Grey granite building (9 metres)
Established: 1900
Admiralty number: A4174
Light characteristic: Fl(3)WR 18s
Grid reference: NM 754 343
Access inside tower: No

Access to site: Mull is accessible by vehicle ferry from Craignure, Lochaline or Kilchoan. From Duart Castle car park it is a short walk of around 1km (0.6 miles) to the east to reach the lighthouse.

7.12 LADY'S ROCK (r)

South west of Lismore

○ Visited on ....

Red clad metal framework tower on white concrete base (12 metres)
Admiralty number: A4172
Light characteristic: Fl W 6s
Grid reference: NM 772 344
Access inside tower: No

Access to site: The lighthouse is visible from ferries from Oban to Mull and the Western Isles. A closer view would require a chartered boat.

7.13 LISMORE (i) ☆

On Eilean Musdile island, off of south eastern point of Lismore

○ Visited on ....

White tower (26 metres)
Established: 1833

Designed by: Robert Stevenson
Admiralty number: A4170
Light characteristic: Fl W 10s
Grid reference: NM 778 351
Access inside tower: No

Access to site: The lighthouse is visible from sailings from Oban to Mull, Coll/Tiree, South Uist and Barra. To access the island a chartered boat is required.

7.14 SGEIR BHUIDHE (t) ☆

○ Visited on ....

Off of coast at Port Appin

White tower on concrete base (8 metres)
Established: 2002
Admiralty number: A4160
Light characteristic: Fl(2)WR 7s
Grid reference: NM 905 462
Access inside tower: No

Access to site: The lighthouse is accessible at low tide by heading north across the beach from Port Appin.

Note: The top section of the old lighthouse is now on display next to Port Appin Village Hall. It features an information display about the original lighthouse and local area.

7.15 DUNOLLIE (m) ☆

○ Visited on ....

North of Oban

Stone tower, white lantern (6 metres)
Established: 1892
Designed by: David A and Charles Stevenson
Admiralty number: A4178
Light characteristic: Fl(2)WRG 6s
Grid reference: NM 852 311
Access inside tower: No

Access to site: The lighthouse is at the side of Ganavan Road, heading north along the coast out of Oban.

7.16 SGEIREAN DUBH (i)

Off of south east coast of Kerrera

○ Visited on ....

White clad metal framework tower, concrete base (7 metres)
Admiralty number: A4186
Light characteristic: Fl(2)W 12s
Grid reference: NM 817 265
Access inside tower: No

Access to site: The lighthouse is visible from the ferry to Colonsay from Oban. A closer view would require a chartered boat.

7.17 DUBH SGEIR (i)

West of Luing

○ Visited on ....

White clad metal framework tower (7 metres)
Established: 2002
Admiralty number: A4192
Light characteristic: Fl WRG 6s
Grid reference: NM 727 120
Access inside tower: No

Access to site: The lighthouse can be seen from the road south of Cullipool on Luing, accessible by vehicle ferry from Cuan. The lighthouse on Fladda (see 7.18) can also be viewed from the area. A closer view would require a chartered boat.

7.18 FLADDA (i) ✯

On Fladda island, north of Sound of Luing

○ Visited on ....

White tower (13 metres)
Established: 1860
Designed by: David and Thomas Stevenson
Admiralty number: A4190
Light characteristic: Fl(2)WRG 9s
Grid reference: NM 721 123
Access inside tower: No

Access to site: Best seen by chartered boat. Landing is on a pier, which leads directly to the lighthouse compound. The lighthouse can be seen from a distance from Cullipool on Luing, which is accessible by vehicle ferry from Cuan. The Dubh Sgeir light (see 7.17) can also be viewed from the area.

Note: The old lens from the lighthouse is now in the Atlantic Islands Centre on Luing.

FLADDA

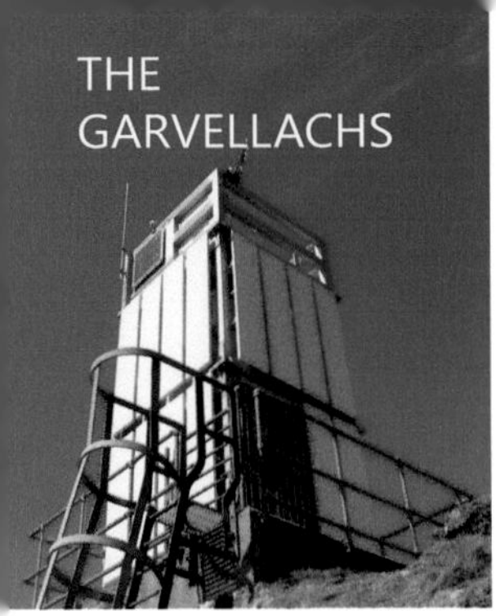

7.19 THE GARVELLACHS (i)

Southern point of Eileach an Naoimh, The Garvellachs

○ Visited on ....

White clad metal framework tower (5 metres)
Established: 2003
Admiralty number: A4196
Light characteristic: Fl(2)W 30s
Grid reference: NM 634 093
Access inside tower: No

Access to site: The lighthouse is visible from the Oban to Colonsay ferry. A closer view would require a chartered boat.

7.20 SCALASAIG (i)

East coast of Colonsay

○ Visited on ....

White clad metal framework tower (5 metres)
Established: 2003
Admiralty number: A4200
Light characteristic: Fl(2)WR 10s
Grid reference: NR 398 939
Access inside tower: No

Access to site: Colonsay is covered by scheduled ferries from Oban. Access to the lighthouse is from The Pantry close to the ferry terminal. A short walk from here across uneven ground takes you to the lighthouse.

Note: Whilst on Colonsay a visit to Colonsay House, where the original lens from Ruvaal lighthouse (see 7.27) on Islay is now a feature in the gardens, may be worthwhile.

7.21 RHINNS OF ISLAY (RINNS OF ISLAY) (i)

On Orsay island, off of Port Wemyss, Islay ☆

○ Visited on ....

White tower (29 metres)
Established: 1825
Designed by: Robert Stevenson
Admiralty number: A4256
Light characteristic: Fl W 5s
Grid reference: NR 164 514
Access inside tower: No

Access to site: The lighthouse is clearly visible from Port Wemyss on Islay with a number of good vantage points along the coastal roads. Access to the island

is by chartered boat only. Once on the island it is a short uphill walk on a path to the lighthouse.

Note: the remains of the old lens from the lighthouse are on display at the Museum of Scottish Lighthouses, Fraserburgh.

7.22 LOCH INDAAL (RUBH'AN DÙIN) (i) ✯

North of Port Charlotte, Islay

○ Visited on ....

White stone tower (13 metres)
Established: 1869
Designed by: David and Thomas Stevenson
Admiralty number: A4254
Light characteristic: Fl(2)WR 7s
Grid reference: NR 257 588
Access inside tower: No

Access to site: Islay is accessible by scheduled vehicle ferry from Kennacraig. The lighthouse can be reached by a short walk north from Port Charlotte, initially following the main road and then through the field to the south of Loch Indaal House.

7.23 CARRAIG FHADA (i-t)

Opposite Port Ellen, south of Islay

○ Visited on ....

White square tower (17 metres)
Established: 1832
Admiralty number: A4248
Light characteristic: Fl WRG 3s
Grid reference: NR 350 444
Access inside tower: No

Access to site: Scheduled vehicle ferries run to Islay from Kennacraig. Once on the island, the light is accessible on foot from the public parking area beyond the cemetery at Kilnaughton Bay, to the west of Port Ellen (signposted Oa Peninsula). The lighthouse is visible from the start of the 1km (0.6 mile) walk. Access to the light may be restricted at the highest tides and in rough sea conditions. The lighthouse can also be seen from the ferry into Port Ellen.

7.24 EILEAN A' CHÙIRN (i) ✯

○ Visited on ....

South east of Islay

White tower (5 metres)
Established: 1907
Designed by: David A and Charles Stevenson
Admiralty number: A4242
Light characteristic: Fl(3)W 18s
Grid reference: NR 473 490
Access inside tower: No

Access to site: By chartered boat. On a clear day it may be possible to see the lighthouse from the ferry to Islay.

7.25 MCARTHUR'S HEAD (i) ✯

○ Visited on ....

East coast of Islay

White tower (13 metres)
Established: 1861
Designed by: David and Thomas Stevenson
Admiralty number: A4240
Light characteristic: Fl(2)WR 10s
Grid reference: NR 462 597
Access inside tower: No

Access to site: Best access is by chartered boat with landing on the jetty and a walk up a number of steps to the lighthouse. It is possible to reach the lighthouse on foot by heading north from Ardtalla on Islay, although this is a difficult walk on very boggy terrain. The lighthouse can be seen from a distance from either the Kennacraig to Port Askaig ferry, or from the main A846 at the south end of Jura.

Note: The old McArthur's Head lens is now located at the exhibition in Mull of Galloway lighthouse (see 10.7 for access details).

EILEAN A' CHÙIRN

MCARTHUR'S HEAD

7.26 CARRAIG MHÒR (i) ✯

○ Visited on ....

South of Port Askaig, Islay

White tower (7 metres)
Established: 1928
Designed by: David A and Charles Stevenson
Admiralty number: A4238
Light characteristic: Fl(2)WR 6s
Grid reference: NR 433 684
Access inside tower: No

Access to site: The light can be seen clearly from the ferry between Kennacraig to Port Askaig on the approach to Islay. It is possible to walk to the lighthouse, though it is not an easy walk. Access on foot is through the gate close to the layby on the road out of Port Askaig. Follow the track south until you reach the small loch, and then continue to the east. As the path begins to head north, leave the path and walk due east across very rough terrain. The walk to the lighthouse from the layby is roughly 1.5kms (1 mile).

7.27 RUVAAL (i) ✯

○ Visited on ....

Northern point of Islay

White tower (34 metres)
Established: 1859
Designed by: David and Thomas Stevenson
Admiralty number: A4236
Light characteristic: Fl(3)W 15s
Grid reference: NR 426 793
Access inside tower: No

Access to site: Islay is accessible by vehicle ferry from Kennacraig. A 6km (3.75 mile) walk north along a track from Bunnahabhain takes you to the lighthouse.

Note: The original lens from Ruvaal lighthouse has been preserved and is now a feature in the gardens at Colonsay House on the nearby island of Colonsay.

7.28 CARRAGH AN T'SRUITH (i) ☆

○ Visited on ....

On west coast of Jura

White tower (5 metres)
Established: 1960
Admiralty number: A4237
Light characteristic: Fl W 3s
Grid reference: NR 439 719
Access inside tower: No

Access to site: Jura is accessible by vehicle ferry from Islay, arriving at Feolin, or by summer only passenger ferry from Tayvallich on the mainland. The lighthouse is accessible on foot from the Feolin ferry terminal by a 3.5km (2 mile) walk north along the coast. The summer only ferry between Colonsay and Islay passes close to the lighthouse. The lighthouse is also visible from a distance on the road to Bunnahabhain on Islay, or at a greater distance from ferries arriving at/departing from Port Askaig on Islay.

7.29 EILEAN NAN GABHAR (i)

○ Visited on ....

East of Craighouse, Jura

White clad metal framework tower (5 metres)
Admiralty number: A4230.6
Light characteristic: Fl W 5s
Grid reference: NR 535 670
Access inside tower: No

Access to site: The lighthouse is visible from the pier at Craighouse, Jura, or from the summer only passenger ferry on the approach to Craighouse. A closer view would require a chartered boat.

7.30 NA CÙILTEAN (i)

○ Visited on ....

South of Sound of Jura

White clad metal framework tower (5 metres)
Established: 2005
Admiralty number: A4232
Light characteristic: Fl W 10s
Grid reference: NR 548 645
Access inside tower: No

Access to site: The lighthouse can be seen at a distance on a clear day from ferries between Kennacraig and Port Askaig on Islay. A closer view would require a chartered boat.

7.31 SKERVUILE (r) ☆

In Sound of Jura

◯ Visited on

....

White tower (25 metres)
Established: 1865
Designed by: David and Thomas Stevenson
Admiralty number: A4230
Light characteristic: Fl W 15s
Grid reference: NR 605 713
Access inside tower: No

Access to site: Best seen by chartered boat, although the Tayvallich to Jura summer only passenger ferry passes close to the lighthouse. The lighthouse is also visible from a distance from the main A846 on Jura. On a clear day it is also possible to see the tower from ferries between Kennacraig and Port Askaig on Islay.

CRINAN CANAL

7.32 RUADH SGEIR (i) ✮

○ Visited on

....

North of Sound of Jura

White tower (7 metres)
Admiralty number: A4226
Light characteristic: Fl W 6s
Grid reference: NR 721 926
Access inside tower: No

Access to site: The lighthouse is difficult to see so is best viewed from a chartered boat only.

7.33 REISA AN T-SRUITH (i)

○ Visited on

....

North east of Jura

White clad metal framework tower (7 metres)
Admiralty number: A4208
Light characteristic: Fl(2)W 12s
Grid reference: NR 734 989
Access inside tower: No

Access to site: The lighthouse is visible from the east coast of Jura via a 14km (8.75 mile) walk north east of Ardlussa along a track. Otherwise a chartered boat would be required.

7.34 CRINAN CANAL (m) ✮

○ Visited on

....

At entrance to Crinan Canal

White tower, red band (6 metres)
Established: 1851
Admiralty number: A4212
Light characteristic: Fl WG 3s
Grid reference: NR 789 944
Access inside tower: No

Access to site: A public car park is available at Crinan sea loch. From here it is a short walk to the lighthouse.

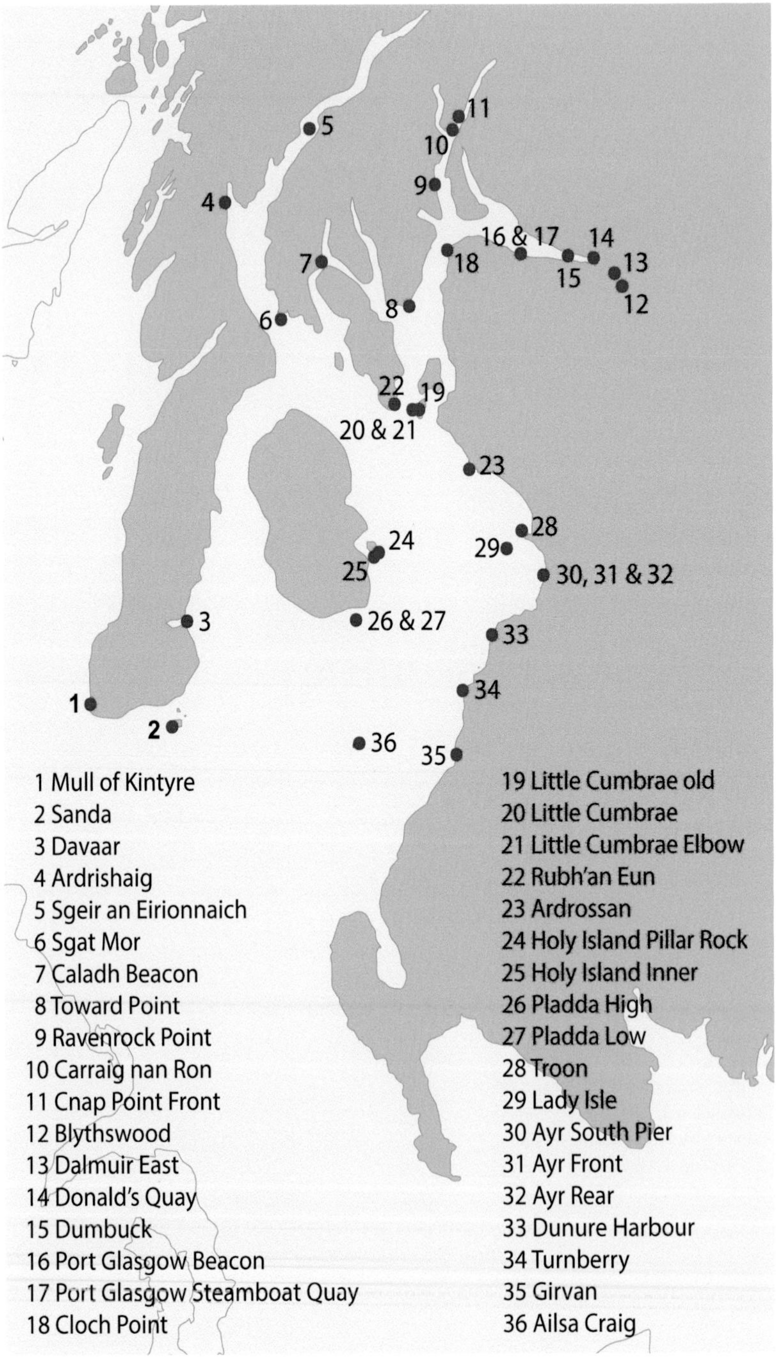
1 Mull of Kintyre
2 Sanda
3 Davaar
4 Ardrishaig
5 Sgeir an Eirionnaich
6 Sgat Mor
7 Caladh Beacon
8 Toward Point
9 Ravenrock Point
10 Carraig nan Ron
11 Cnap Point Front
12 Blythswood
13 Dalmuir East
14 Donald's Quay
15 Dumbuck
16 Port Glasgow Beacon
17 Port Glasgow Steamboat Quay
18 Cloch Point
19 Little Cumbrae old
20 Little Cumbrae
21 Little Cumbrae Elbow
22 Rubh'an Eun
23 Ardrossan
24 Holy Island Pillar Rock
25 Holy Island Inner
26 Pladda High
27 Pladda Low
28 Troon
29 Lady Isle
30 Ayr South Pier
31 Ayr Front
32 Ayr Rear
33 Dunure Harbour
34 Turnberry
35 Girvan
36 Ailsa Craig

Region 8

Kintyre And Firth Of Clyde

8.1 MULL OF KINTYRE (m) ☆

South eastern point of Kintyre

White tower (12 metres)
Established: 1820s – replacing 1788 Thomas Smith tower
Admiralty number: A4272
Light characteristic: Fl(2)W 20s
Grid reference: NR 588 084
Access inside tower: No

○ Visited on ....

Access to site: No vehicles are permitted on the access road, but parking is available at the end of the public road beyond Southend. From here the lighthouse is accessible by a 4km (2.5 mile) return walk on a very steep tarmac road. On a clear day is it possible to see Sanda island and lighthouse (see 8.2) from the road before you reach the parking area.

8.2 SANDA (i) ✯

At south of Sanda island, off of south east coast of Kintyre peninsula

White tower (15 metres)
Established: 1850
Designed by: Alan Stevenson
Admiralty number: A4274
Light characteristic: Fl W 10s
Grid reference: NR 726 038
Access inside tower: No

○ Visited on

....

Access to site: The island is privately owned. Due to the location of the lighthouse on the island, it is difficult to see from any distance, although it can be seen on a clear day from the road to the Mull of Kintyre. Best seen by chartered boat.

Note: The old lens and mechanism from the lighthouse is now on display at the Museum of Scottish Lighthouses, Fraserburgh.

8.3 DAVAAR (t) ✯

East of Campbeltown

○ Visited on

....

White tower (20 metres)
Established: 1854
Designed by: David and Thomas Stevenson
Admiralty number: A4276
Light characteristic: Fl(2)W 10s
Grid reference: NR 761 207
Access inside tower: No

Access to site: Access to the island and lighthouse is only possible at low tide and tide times should be checked before travelling to allow enough time to get both to and from the island. From Campbeltown follow the coastal road to the east. Access to the beach is through a gate and there is parking a little further on. From the gate it is a 2km (1.25 mile) walk to the lighthouse across a sand/pebble causeway. Once on the island follow the path around the north of the island to reach the lighthouse.

Note: The former parabolic mirrors from the lighthouse are now on display at the Museum of Scottish Lighthouses, Fraserburgh.

8.4 ARDRISHAIG (m) ✯

At end of breakwater, Ardrishaig

◯ Visited on ....

White tower (6 metres)
Admiralty number: A4304
Light characteristic: LFl WRG 6s
Grid reference: NR 855 854
Access inside tower: No

Access to site: The breakwater is open to the public. Parking is available off of Chalmers Street. From here it is a short walk over the bridge and along to the end of the breakwater.

8.5 SGEIR AN EIRIONNAICH (PADDY ROCK, LOCH FYNE) (i)

In Loch Fyne

◯ Visited on ....

Black metal framework tower on black round tower, white stripes (7 metres)
Admiralty number: A4314
Light characteristic: Fl WR 3s
Grid reference: NR 995 953
Access inside tower: No

Access to site: The light is visible from a distance from a parking area north of Barnacarry, or from further away on the A83 at Minard. A chartered boat would be required for a closer view.

8.6 SGAT MÓR (i) ✯

South end of Sgat Mór island, south of entrance to Asgog Bay

◯ Visited on ....

White round tower on concrete base (8 metres)
Admiralty number: A4296
Light characteristic: Fl W 3s
Grid reference: NR 930 665
Access inside tower: No

Access to site: Best seen by chartered boat. The lighthouse can be seen from the mainland beyond the end of a track south from the main road towards Portvadie. The walk from the main road is 2.5kms (1.5 miles) to the coast where the lighthouse is visible.

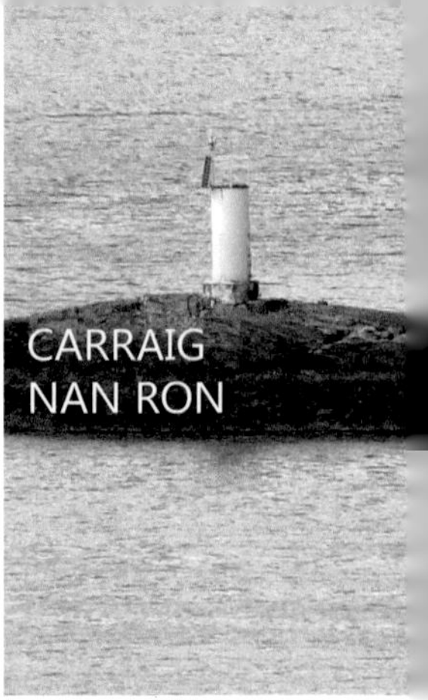

8.7 CALADH BEACON (m)

○ Visited on ....

At west entrance to Loch Riddon

Masonry tower (8 metres)
Established: 1888
Light discontinued
Grid reference: NS 005 756
Access inside tower: No

Access to site: The lighthouse is accessible on foot from Tighnabruaich where public parking is available. From here it is a 4km (2.5 mile) walk north east along the coastal road to reach the lighthouse.

8.8 TOWARD POINT (m) ✯

○ Visited on ....

In Toward village

White tower (19 metres)
Established: 1812
Designed by: Robert Stevenson
Admiralty number: A4362
Light characteristic: Fl W 10s
Grid reference: NS 136 673
Access inside tower: No

Access to site: The site is privately owned. The turn-off to the lighthouse is signposted from the main A815. Access to the site is along the private road between Post House and Pier Cottage.

8.9 RAVENROCK POINT (m)

○ Visited on ....

West bank of Loch Long

White metal framework tower on white column (8 metres)
Admiralty number: A4409
Light characteristic: Fl W 4s
Grid reference: NS 190 863
Access inside tower: No

Access to site: The light is at the side of the coastal road, around 1km (0.6 miles) south of Ardentinny.

8.10 CARRAIG NAN RON (DOG ROCK) (i)

○ Visited on ....

Off of southern point of Argyll Forest Park

White column (5 metres)
Admiralty number: A4411.4
Light characteristic: Fl W 2s
Grid reference: NS 221 934
Access inside tower: No

Access to site: The light is visible from a distance from Feuins Road, Portincaple. Best seen from a chartered boat.

8.11 CNAP POINT FRONT (m)

○ Visited on ....

West bank of Loch Long

White column (5 metres)
Admiralty number: A4412
Light characteristic: Q W
Grid reference: NS 240 959
Access inside tower: No

Access to site: The light is visible from a distance from A814 south of Glenmallan. A closer view would require a chartered boat.

8.12 BLYTHSWOOD (m) ✯

○ Visited on ....

North of Renfrew

Green tower, green base (11 metres)
Established: 1921
Admiralty number: A4512
Light characteristic: Fl G 2s
Grid reference: NS 499 690
Access inside tower: No

Access to site: The lighthouse is on the Clyde Coastal Path, north of Renfrew Golf Club.

8.13 DALMUIR EAST (m)

○ Visited on ....

South of Clydebank

Red framework tower, red base (11 metres)
Admiralty number: A4510
Light characteristic: Fl R 4s
Grid reference: NS 490 704
Access inside tower: No

Access to site: The light is visible from a distance from Cable Depot Road, Dalmuir. A closer view would require a chartered boat.

BLYTHSWOOD

DALMUIR EAST

8.14 DONALD'S QUAY (m)

○ Visited on ....

North west of Erskine Bridge

Red framework tower on red hut (11 metres)
Admiralty number: A4496
Light characteristic: Fl R 4s
Grid reference: NS 459 730
Access inside tower: No

Access to site: The light is near the canal crossing on Portpatrick Road. A small path opposite the crossing takes you through the trees to the lighthouse.

8.15 DUMBUCK (r)

○ Visited on ....

In River Clyde, south of Dumbarton

Green masonry tower (12 metres)
Established: 1921
Admiralty number: A4486
Light characteristic: Fl G 2s
Grid reference: NS 413 737
Access inside tower: No

Access to site: The lighthouse can be seen from the end of Castle Road near Dumbarton Castle. A closer view would require a chartered boat.

8.16 PORT GLASGOW BEACON (r)

○ Visited on ....

Off of coast at Port Glasgow ☆

Black and white checkered tower and cupola (8 metres)
Admiralty number: A4476
Light characteristic: F G
Grid reference: NS 322 749
Access inside tower: No

Access to site: The lighthouse is best seen from Mirren's Shore near the Port Glasgow Steamboat Quay light (see 8.17).

8.17 PORT GLASGOW STEAMBOAT QUAY (m) ☆

○ Visited on ....

In Port Glasgow

Black and white checkered column (11 metres)

DUMBUCK

PORT GLASGOW BEACON

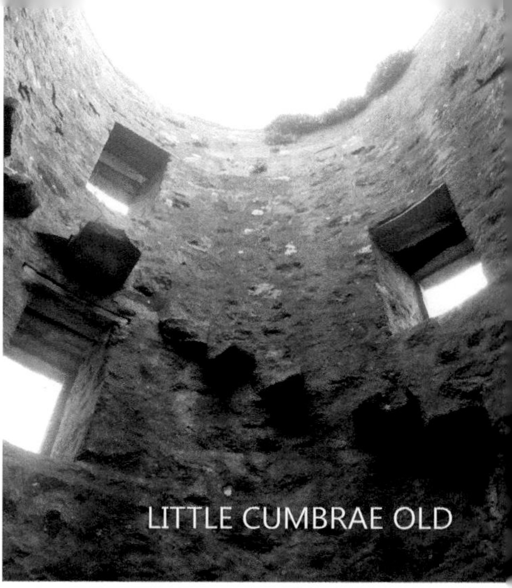

Admiralty number: A4478
Light characteristic: F G
Grid reference: NS 320 749
Access inside tower: No

Access to site: The lighthouse is accessible at the north west end of Mirren's Shore. The Port Glasgow Beacon (see 8.16) can also be seen clearly from here.

8.18 CLOCH POINT (m) ☆

South west of Gourock

○ Visited on

....

White round tower, black band (24 metres)
Established: 1797
Designed by: Thomas Smith
Admiralty number: A4404
Light characteristic: Fl W 3s
Grid reference: NS 203 759
Access inside tower: No

Access to site: The lighthouse is at the side of A770 at Cloch.

8.19 LITTLE CUMBRAE OLD (i)

Near high point of Little Cumbrae

○ Visited on

....

Round stone tower
Established: 1757
Light discontinued 1793
Grid reference: NS 144 515
Access inside tower: Access to ground level.

Access to site: Boat trips operating out of Largs can be arranged to reach Little Cumbrae. The tower is accessible by a 1.5km (1 mile) walk along a path from east landing. The final section is across rougher ground. From the west landing, beyond the Little Cumbrae lighthouse (see 8.20) and Cumbrae Elbow light (see 8.21) it is a 1km (0.6 mile) walk along the path to the south east before heading north on uneven ground to reach the tower.

8.20 LITTLE CUMBRAE (i) ☆

On western point of Little Cumbrae

○ Visited on
....

Octagonal stone tower (11 metres)
Established: 1793
Designed by: Thomas Smith and Robert Stevenson
Light discontinued 1997
Grid reference: NS 138 515

Access inside tower: Tower and associated accommodation lie derelict and open.

Access to site: Boat trips operating out of Largs can be arranged to reach Little Cumbrae. From the east landing it is a 2.5km (1.5 mile) walk to the lighthouse, via the old lighthouse. From the west landing it is a short uphill walk to the lighthouse. The Cumbrae Elbow light (see 8.21) is a short distance away. From here the old tower (see 8.19) is a 1km (0.6 mile) walk to the south east along the path before heading north across rougher ground.

8.21 CUMBRAE ELBOW (i)

On west side of Little Cumbrae

○ Visited on
....

White tower (11 metres)
Established: 1997
Admiralty number: A4346
Light characteristic: Fl W 6s
Grid reference: NS 137 514
Access inside tower: No

Access to site: Boat trips operating out of Largs can be arranged to reach Little Cumbrae. From the east landing it is a 2.5km (1.5 mile) walk to the lighthouse, via the old lighthouse (see 8.19). From the west landing it is a short uphill walk to the lighthouse. The Little Cumbrae lighthouse (see 8.20) is a short distance away. From here the old tower (see 8.19) is a 1km (0.6 mile) walk to the south east along the path before heading north across rougher ground.

8.22 RUBH'AN EUN (i) ☆

South east point of Isle of Bute

○ Visited on
....

White metal tower (8 metres)
Established: 1911
Admiralty number: A4344

Light characteristic: Fl R 6s
Grid reference: NS 115 526
Access inside tower: No

Access to site: Scheduled vehicle ferries run from Wemyss Bay to Rothesay on the Isle of Bute. The A844 takes you around the coast and a turn off at Kingarth takes you towards Kilchattan Bay. There is a parking area at Kilchattan Bay. From here head south and follow the coast for 2kms (1.25 miles) which will take you to the lighthouse. The track is very narrow places.

8.23 ARDROSSAN LIGHTHOUSE PIER (m) ☆

At end of west pier, Ardrossan

○ Visited on ....

White tower (9 metres)
Established: 1898
Admiralty number: A4538
Light characteristic: Iso WG 4s
Grid reference: NS 222 422
Access inside tower: No

Access to site: The pier is accessible on foot from the end of Harbour Road near the ferry terminal.

8.24 HOLY ISLAND PILLAR ROCK POINT (HOLY ISLAND OUTER) (i) ☆

East coast of Holy Island

○ Visited on ....

White square tower (23 metres)
Established: 1905
Designed by: David A and Charles Stevenson
Admiralty number: A4330
Light characteristic: Fl(2)W 20s
Grid reference: NS 069 292
Access inside tower: No

Access to site: Regular boat trips run to Holy Island from Lamlash on Arran (accessible by scheduled ferry from Ardrossan). From the jetty on Holy Island a 3.5km (2 mile) walk following the coastal path around the south of the island takes you to the lighthouse, passing the Holy Island Inner lighthouse (see 8.25).

8.25 HOLY ISLAND INNER (i) ☆

○ Visited on ....

South west point of Holy Island

White tower (17 metres)
Established: 1877
Designed by: David and Thomas Stevenson
Admiralty number: A4332
Light characteristic: Fl G 3s
Grid reference: NS 062 286
Access inside tower: No

Access to site: Regular boat trips run to Holy Island from Lamlash on Arran (accessible by scheduled ferry from Ardrossan). From the jetty on Holy Island a 2.5km (1.5 mile) walk following the coastal path to the south of the island takes you to the lighthouse, which is in a private retreat with no visitors allowed. A further 1km (0.6 miles) around the coast takes you to the Pillar Rock Point lighthouse (see 8.24).

8.26 PLADDA HIGH (i) ☆

○ Visited on ....

South of Isle of Arran

White tower (29 metres)
Established: 1790
Designed by: Thomas Smith
Admiralty number: A4326
Light characteristic: Fl(3)W 30s
Grid reference: NS 027 190
Access inside tower: No

Access to site: The lighthouse can be seen from Arran, where it is also possible to book a boat trip to see the island. The Pladda low light (see 8.27) is within the same building.

8.27 PLADDA LOW (i) ☆

○ Visited on ....

South of Isle of Arran

White tower (13 metres)
Established: 1790

Designed by: Thomas Smith
Light discontinued 1901
Grid reference: NS 027 190
Access inside tower: No

Access to site: The lighthouse can be seen from Arran, where it is also possible to book a boat trip to see the island. The Pladda high light (see 8.26) is within the same building.

8.28 TROON HARBOUR (m) ☆

○ Visited on ....

At end of west pier, Troon

White tower (8 metres)
Established: 1904
Admiralty number: A4554
Light characteristic: Fl(2)WG 5s
Grid reference: NS 308 319
Access inside tower: No

Access to site: There is no unauthorised access to the pier. The lighthouse can be viewed from the end of the public road at the entrance to the pier.

Note: The previous lantern from the lighthouse is now located outside the Scottish Maritime Museum in Irvine.

8.29 LADY ISLE (i) ☆

○ Visited on ....

South west of coast of Troon

White tower, red stripes (5 metres)
Established: 1903
Designed by: David A and Charles Stevenson
Admiralty number: A4562
Light characteristic: Fl W 2s
Grid reference: NS 275 294
Access inside tower: No

Access to site: The island is accessible by chartered boat only. On a clear day the tower can be seen from Troon seafront.

LADY ISLE

8.30 AYR SOUTH PIER (m)

At end of south pier, Ayr

○ Visited on ....

White tower (6 metres)
Established: 1877
Admiralty number: A4568
Light characteristic: Q W
Grid reference: NS 329 227
Access inside tower: No

Access to site: The pier is accessible at the end of the Esplanade. A small parking area is at the entrance to the pier. The Ayr Front (see 8.31) and Ayr Rear (see 8.32) lights can be seen from the pier.

8.31 AYR FRONT (m)

At entrance to Griffin Dock, Ayr

○ Visited on ....

Red tower, white lantern (6 metres)
Established: 1841
Admiralty number: A4574
Light characteristic: F R
Grid reference: NS 332 227
Access inside tower: No

Access to site: There is no unauthorised access to site. The lighthouse is best viewed from South Harbour Street on the other side of the river. The rear light (see 8.32) can also be seen from here. The light can also be seen from the South Pier light (see 8.30).

8.32 AYR REAR (m) ✯

North side of Ayr harbour

○ Visited on ....

Masonry tower, white lantern (22 metres)
Established: 1841
Admiralty number: A4574.1
Light characteristic: Oc R 10s
Grid reference: NS 333 226
Access inside tower: No

Access to site: There is no unauthorised access to site. The lighthouse is best viewed from South Harbour Street on the other side of the river. The front light (see 8.31) can also be seen from here. The light can also be seen from the South Pier light (see 8.30).

8.33 DUNURE HARBOUR (m)

○ Visited on ....

At end of west pier, Dunure

Round sandstone tower (4 metres)
Established: 1811
Light discontinued around 1850
Grid reference: NS 254 160

Access inside tower: A sign at the lighthouse advises against accessing the building due to structural instability.

Access to site: The pier is accessible from the end of Harbour View. A small parking area is at the entrance to the pier.

8.34 TURNBERRY (m) ☆

○ Visited on ....

In grounds of Turnberry golf course

White tower (24 metres)
Established: 1873
Designed by: David and Thomas Stevenson
Admiralty number: A4580
Light characteristic: Fl W 15s
Grid reference: NS 196 073
Access inside tower: No

Access to site: The lighthouse is accessible on foot through Turnberry Golf Course.

Note: The old parabolic mirrors from the lighthouse are now on display at the Museum of Scottish Lighthouses, Fraserburgh.

8.35 GIRVAN (m)

○ Visited on ....

At end of south pier, Girvan

Metal mast on white tower (6 metres)
Admiralty number: A4588

DUNURE HARBOUR

Light characteristic: 2 F G(vert)
Grid reference: NX 180 983
Access inside tower: No

Access to site: The pier is accessible on foot. Parking is at the end of Knockcushan Street.

8.36 AILSA CRAIG (i) ✯

○ Visited on ....

On eastern point of Ailsa Craig island, off of coast of Girvan

White tower (11 metres)
Established: 1886
Designed by: Thomas and David A Stevenson
Admiralty number: A4582
Light characteristic: Fl W 4s
Grid reference: NX 026 997
Access inside tower: No

Access to site: Regular sailings operate from Girvan to Ailsa Craig throughout the year. Once landed on the island, a short walk takes you to the lighthouse.

Note: The former lens and mechanism from the lighthouse is now on display in the Museum of Scottish Lighthouses, Fraserburgh.

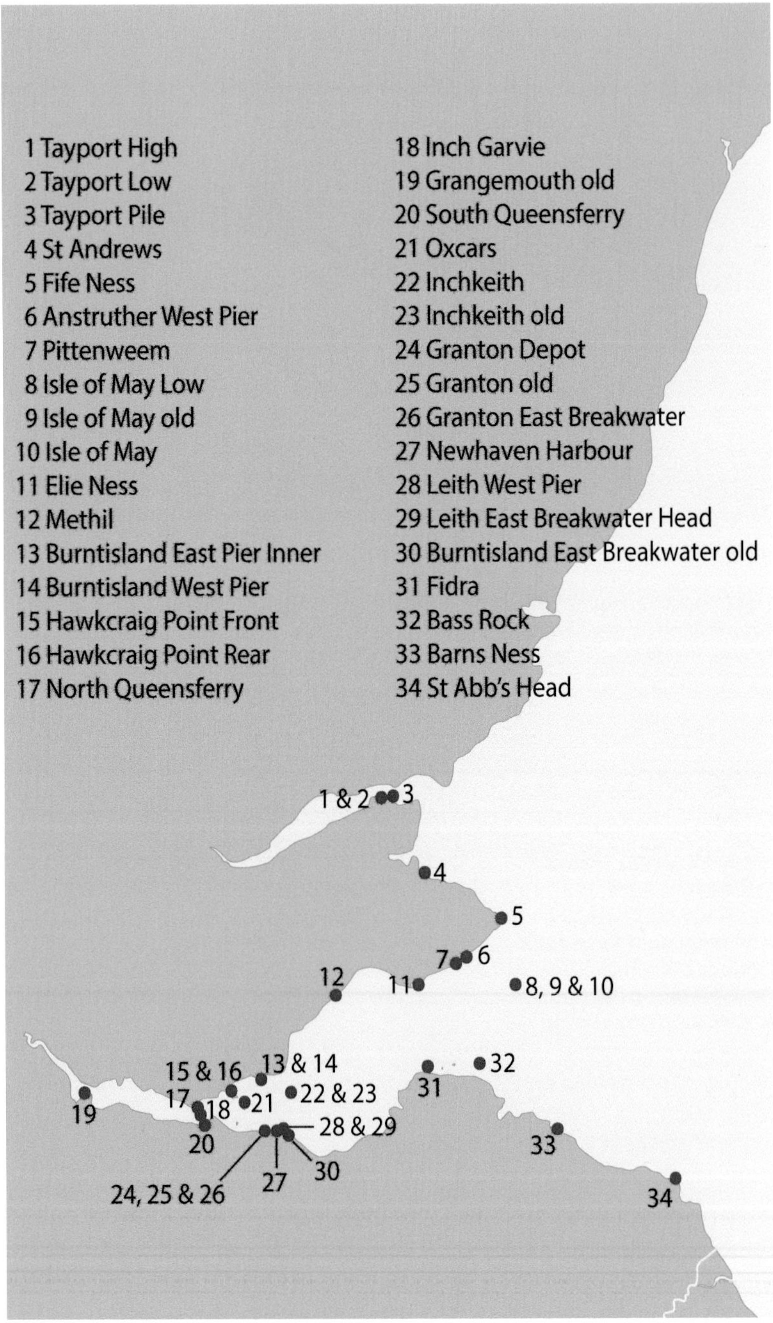
1 Tayport High
2 Tayport Low
3 Tayport Pile
4 St Andrews
5 Fife Ness
6 Anstruther West Pier
7 Pittenweem
8 Isle of May Low
9 Isle of May old
10 Isle of May
11 Elie Ness
12 Methil
13 Burntisland East Pier Inner
14 Burntisland West Pier
15 Hawkcraig Point Front
16 Hawkcraig Point Rear
17 North Queensferry
18 Inch Garvie
19 Grangemouth old
20 South Queensferry
21 Oxcars
22 Inchkeith
23 Inchkeith old
24 Granton Depot
25 Granton old
26 Granton East Breakwater
27 Newhaven Harbour
28 Leith West Pier
29 Leith East Breakwater Head
30 Burntisland East Breakwater old
31 Fidra
32 Bass Rock
33 Barns Ness
34 St Abb's Head
1 & 2
3
4
5
7
6
12
11
8, 9 & 10
15 & 16
13 & 14
32
31
17
18
21
22 & 23
19
20
28 & 29
33
30
27
24, 25 & 26
34

Region 9

South East Scotland

9.1 TAYPORT HIGH (TAYPORT WEST) (m) ☆

West of Tayport

White tower (23 metres)
Established: 1823
Designed by: Robert Stevenson
Admiralty number: A3142
Light characteristic: Dir Iso WRG 3s
Grid reference: NO 447 294
Access inside tower: No

○ Visited on ....

Access to site: Public parking is available off of Commonty Road. From here it is a short walk west to the lighthouse, beyond the low lighthouse (see 9.2).

9.2 TAYPORT LOW (TAYPORT EAST) (m) ✯

West of Tayport

◯ Visited on ....

Stone tower (12 metres)
Established: 1823
Designed by: Robert Stevenson
Light discontinued 1848
Grid reference: NO 451 293
Access inside tower: No

Access to site: Public parking is available off of Commonty Road. From here it is a short walk west to the lighthouse. The high lighthouse (see 9.1) is a little further along the path.

9.3 TAYPORT PILE (LARIK BEACON) (r)

East of Tayport

◯ Visited on ....

Lantern and wooden octagonal structure on wood pilings
Established: 1848
Light discontinued 1960s
Grid reference: NO 463 293
Access inside tower: No

Access to site: The light is visible from a distance from Harbour Road, just before the entrance to the harbour.

9.4 ST ANDREWS (m)

On east side of St Andrews Cathedral ruins, above harbour

◯ Visited on ....

Semi-circular stone tower (5 metres)
Light discontinued
Grid reference: NO 515 166
Access inside tower: No

Access to site: Parking is available on Shorehead near the entrance to the north pier. The lighthouse is visible from here and is accessible via a short uphill walk towards the cathedral.

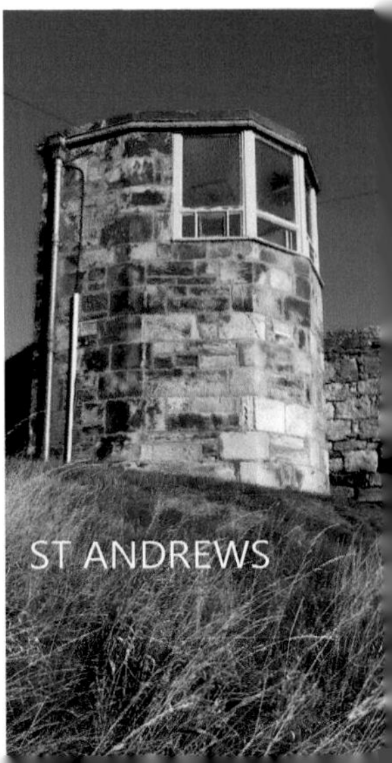

9.5 FIFE NESS (m) ✮

In grounds of Craighead Links, north of Crail

◯ Visited on ....

White building (5 metres)
Established: 1975
Designed by: Peter Hyslop
Admiralty number: A3102
Light characteristic: Iso WR 10s
Grid reference: NO 638 098
Access inside tower: No

Access to site: From the car park at Craighead Links it is a short walk along the road to the lighthouse.

9.6 ANSTRUTHER WEST PIER (CHALMERS) (m)

At end of west pier, Anstruther ✮

◯ Visited on ....

White octagonal tower (9 metres)
Established: 1880
Light discontinued
Grid reference: NO 569 032
Access inside tower: No

Access to site: The pier is accessible on foot. Public parking is available on Shore Street near the pier entrance.

9.7 PITTENWEEM EAST PIER (m)

Halfway along east pier, Pittenweem

◯ Visited on ....

Round stone tower (5 metres)
Light discontinued 1995
Grid reference: NO 550 023
Access inside tower: No

Access to site: The pier is accessible on foot. Parking is available on East Shore, a short walk away.

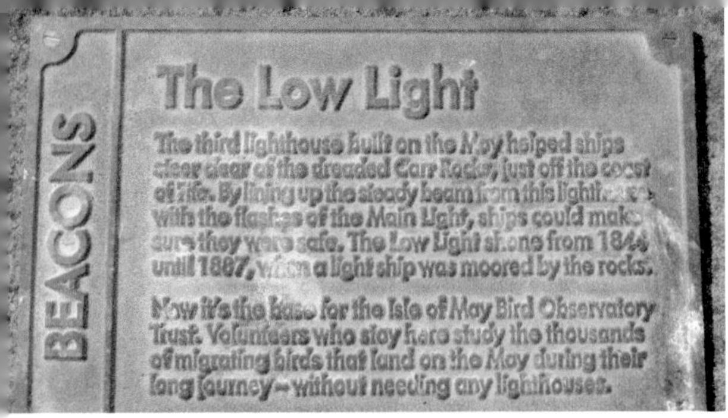

9.8 ISLE OF MAY LOW (i) ✯

East coast of Isle of May

○ Visited on ....

White round tower (18 metres)
Established: 1844
Designed by: Alan Stevenson
Light discontinued 1887
Grid reference: NT 655 996
Access inside tower: No

Access to site: Regular boat trips run from Anstruther to the Isle of May throughout the summer months. The lighthouse is a short walk from the active lighthouse (see 9.10) and the old light (see 9.9).

9.9 ISLE OF MAY OLD (i)

Close to Isle of May lighthouse

○ Visited on ....

White square castellated stone tower (12 metres)
Established: 1636
Light discontinued 1816
Grid reference: NT 656 994
Access inside tower: No

Access to site: Regular boat trips run from Anstruther to the Isle of May throughout the summer months. The building can be seen close to the active lighthouse (see 9.10), and the low light (see 9.8) is a short walk away.

9.10 ISLE OF MAY (i) ✯

On highest point of Isle of May

○ Visited on ....

Square tower on stone dwelling (24 metres)
Established: 1816
Designed by: Robert Stevenson
Admiralty number: A3090
Light characteristic: Fl(2)W 15s
Grid reference: NT 655 994

Access inside tower: On special occasions it is possible to access the tower.

Access to site: Regular boat trips run from Anstruther to the Isle of May throughout the summer months. It is a short uphill walk from the pier. The old light (see 9.9) is nearby and the low light (see 9.8) is a short walk away.

Note: The old lens from the lighthouse is now on display in the Museum of Scottish Lighthouses, Fraserburgh.

9.11 ELIE NESS (m)

South coast of Elie

◯ Visited on ....

White tower (11 metres)
Established: 1908
Designed by: David A and Charles Stevenson
Admiralty number: A3060
Light characteristic: Fl W 6s
Grid reference: NT 496 993
Access inside tower: No

Access to site: A 1km (0.6 mile) walk from the car park at Ruby Bay takes you along a clearly marked track to the lighthouse.

9.12 METHIL (m) ☆

Entrance to Methil Docks

◯ Visited on ....

White tower (8 metres)
Admiralty number: A3050
Light characteristic: Oc G 6s
Grid reference: NT 375 990
Access inside tower: No

Access to site: The light is accessible on foot from the end of Harbour View.

9.13 BURNTISLAND EAST PIER INNER (m)

In Burntisland Docks

◯ Visited on ....

Rusty white tower (7 metres)
Admiralty number: A3032
Light characteristic: 2 F G(vert)
Grid reference: NT 230 854
Access inside tower: No

Access to site: Access to the lighthouse is near Burntisland train station.

Note: For details of the former lighthouse on the east pier see 9.30.

9.14 BURNTISLAND WEST PIER HEAD (m) ☆

At end of west pier head, Burntisland

◯ Visited on ....

White tower (6 metres)
Established: 1879
Admiralty number: A3026

Light characteristic: Fl(2)R 6s
Grid reference: NT 230 852
Access inside tower: No

Access to lighthouse: The light is visible from a distance from Burntisland train station.

9.15 HAWKCRAIG POINT FRONT (m) ○ Visited on ....

East of Aberdour

White tower, red band (4 metres)
Admiralty number: A2915
Light characteristic: Iso W 5s
Grid reference: NT 201 849
Access inside tower: No

Access to site: The light is at the end of Hawkcraig Road. Parking is available close to lighthouse. The rear light (see 9.16) is to the west.

9.16 HAWKCRAIG POINT REAR (m) ○ Visited on ....

East of Aberdour

White tower, red band (8 metres)
Admiralty number: A2915.1
Light characteristic: Iso W 5s
Grid reference: NT 200 849
Access inside tower: No

Access to site: The light is at the end of Hawkcraig Road. Parking is available close to lighthouse. The front light (see 9.15) is to the east.

9.17 NORTH QUEENSFERRY (m) ☆ ○ Visited on ....

At Town Pier, North Queensferry

Hexagonal stone tower (5 metres)
Established: 1812
Light discontinued

Grid reference: NT 131 804

Access inside tower: Yes. The tower is open throughout the year, weather permitting, with the opportunity to light the lamp.

Access to site: The lighthouse is at the end of the B981, Main Street.

9.18 INCH GARVIE (m) ☆

○ Visited on ....

West side of Forth Road Bridge

Black round beacon, white lantern (10 metres)
Established: 1886
Admiralty number: A2920
Light characteristic: Lfl W 5s
Grid reference: NT 135 796
Access inside tower: No

Access to site: The light can be seen from a distance from North Queensferry lighthouse (see 9.17), but best seem from a boat as you pass under the Forth Rail Bridge. Regular boat trips run from South Queensferry to nearby islands during the summer months.

Note: The Beamer Rock lighthouse, which was located a short distance from here, was removed to allow for the building of the Queensferry crossing and the old lighthouse is currently in storage in the area. It is hoped that, at some point, it will be re-built in the area.

9.19 GRANGEMOUTH OLD (m)

○ Visited on ....

Remains north of Grangemouth

Round masonry tower (originally 9.5 metres, 5 metres remain)
Established: 1876
Light discontinued around 1950
Grid reference: NS 945 838
Access inside tower: No

Access to site: The lighthouse is on a large area of private land. As a result, the remains are only visible from chartered boat.

INCH GARVIE

9.20 SOUTH QUEENSFERRY (m) ✯

At entrance to Hawes Pier next to Forth Rail Bridge, South Queensferry

○ Visited on ....

Hexagonal stone tower (5 metres)
Established: 1812
Light discontinued around 1920
Grid reference: NT 137 784
Access inside tower: No

Access to site: The lighthouse is behind the RNLI South Queensferry building, next to the Forth Rail Bridge. Parking is off of Newhalls Road.

9.21 OXCARS (r) ✯

South of Aberdour, in Firth of Forth

○ Visited on ....

White tower, red band (22 metres)
Established: 1886
Designed by: Thomas and David A Stevenson
Admiralty number: A2916
Light characteristic: Fl(2)WR 7s
Grid reference: NT 203 818
Access inside tower: No

Access to site: The lighthouse is visible from routine boat crossings to nearby islands from South Queensferry. A closer view would require a chartered boat.

9.22 INCHKEITH (i) ✯

On summit of Inchkeith island, in Firth of Forth

○ Visited on ....

Stone tower (19 metres)
Established: 1804
Designed by: Thomas Smith
Admiralty number: A2912
Light characteristic: Fl W 15s
Grid reference: NT 293 829
Access inside tower: No

Access to site: The island is accessible by chartered boat only. Depending on the tide state, landing may involve a climb up a ladder. The lighthouse is visible from Leith to the south or Pettycur to the north. The remains of the old lighthouse (see 9.23) are 100 metres to the south.

Note: The former lens from the lighthouse is now on display at the National Museum of Scotland, Edinburgh.

9.23 INCHKEITH OLD (i)

○ Visited on ....

On Inchkeith island, in Firth of Forth

Circular brick remains with terracotta floor tiles
Established: circa 1785
Designed by: Thomas Smith
Grid reference: NT 293 828
Access inside tower: Yes

Access to site: The island is accessible by chartered boat only. Depending on the tide state, landing may involve a climb up a ladder. The operational lighthouse (see 9.22) is 100 metres to the north.

Note: The light was never routinely lit. The tower was built to test lights prior to their transfer elsewhere.

9.24 GRANTON DEPOT (m) ✯

○ Visited on ....

South of Granton harbour

Brick tower with black lantern on dwellings (15 metres)
Established: 1884
Grid reference: NT 234 772
Access inside tower: On special occasions only.

Access to site: The lighthouse is on West Harbour Street.

Note: The lighthouse was never used as an aid to navigation. It was built for Northern Lighthouse Board training and experiments.

9.25 GRANTON OLD (m)

○ Visited on ....

At end of middle pier, Granton

Steel post with lantern and gallery (12 metres)
Light discontinued 1970
Grid reference: NT 238 776
Access inside tower: No

Access to site: The lighthouse is on private land but can be seen from the end of Lochinvar Drive. Alternatively, the light can be seen from the east breakwater light (see 9.26).

9.26 GRANTON HARBOUR EAST BREAKWATER (m)

○ Visited on ....

At end of east breakwater, Granton

White square brick building (5 metres)
Grid reference: NT 240 779
Access inside tower: No

Access to site: The east breakwater is accessible from the A901, Lower Granton Road. The old lighthouse (see 9.25) is visible from the breakwater.

9.27 NEWHAVEN HARBOUR (m) ✯

○ Visited on ....

On north side of harbour, Newhaven

White cast iron tower (15 metres)
Established: 1869
Light discontinued 1930, although it is now lit for decorative purposes only.
Grid reference: NT 254 772
Access inside tower: No

Access to site: The breakwater is accessible on foot. Car parking is available off of Newhaven Place.

9.28 LEITH WEST PIER (m) ✯

○ Visited on ....

At end of west pier, Leith

White octagonal tower (6 metres)
Established: 1938
Light discontinued 2008
Grid reference: NT 262 780
Access inside tower: No

Access to site: The lighthouse is accessible on foot from the end of Western Harbour View where parking is available. From here it is a short walk through the park and along the pier. The east breakwater lighthouse (see 9.29) is visible from here.

GRANTON HARBOUR
EAST BREAKWATER

NEWHAVEN
HARBOUR

LEITH WEST PIER

9.29 LEITH EAST BREAKWATER HEAD (m) ☆

At end of east breakwater, Leith

◯ Visited on ....

Red lantern on concrete base (4 metres)
Established: 1937
Admiralty number: A2884
Light characteristic: Iso R 4s
Grid reference: NT 263 782
Access inside tower: No

Access to site: The lighthouse is on private land and is best viewed from the Leith West Pier lighthouse (see 9.28).

9.30 BURNTISLAND EAST BREAKWATER OLD (m) ☆

◯ Visited on ....

Relocated to Leith Inner Harbour from Burntisland

White octagonal metal tower, round lantern (7 metres)
Established: 1876

LEITH EAST BREAKWATER HEAD

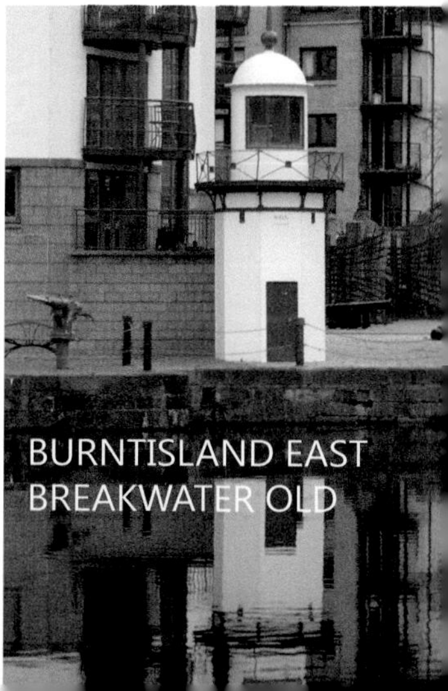

BURNTISLAND EAST BREAKWATER OLD

Light discontinued after 1954
Grid reference: NT 271 767
Access inside tower: No

Access to site: The lighthouse is on the corner of The Shore and Timber Bush on the east bank of the Water of Leith.

9.31 FIDRA (i) ☆

◯ Visited on ....

Near summit of Fidra island, in Firth of Forth

White brick tower (17 metres)
Established: 1885
Designed by: Thomas and David A Stevenson
Admiralty number: A2868
Light characteristic: Fl(4)W 30s
Grid reference: NT 513 870
Access inside tower: No

Access to site: The island is only accessible by chartered boat. Once on the pier, a short uphill walk on an old track takes you to the lighthouse. The lighthouse can be seen from Yellowcraig beach to the south.

9.32 BASS ROCK (i) ☆

◯ Visited on ....

On south side of Bass Rock, in Firth of Forth

White tower (20 metres)
Established: 1903
Designed by: David A and Charles Stevenson
Admiralty number: A2864
Light characteristic: Fl(3)W 20s
Grid reference: NT 602 873
Access inside tower: No

Access to site: The lighthouse is visible from North Berwick and a particularly good viewpoint is close to Tantallon Castle. A closer view can be achieved

from the sea by regular boat trips operating out of North Berwick during the summer months. Landing on the island is possible from occasional trips or chartered boat but requires calm sea conditions.

Note: The old lens from the lighthouse is now on display at the Coastal Communities Museum in North Berwick.

9.33 BARNS NESS (m) ✮

South of Dunbar

○ Visited on ....

White round tower (37 metres)
Established: 1901
Designed by: David A and Charles Stevenson
Light discontinued in 2005
Grid reference: NT 723 772
Access inside tower: No

Access to site: The lighthouse is accessible by road from a turn off of the A187 south of Broxburn. Parking is available a short distance from the lighthouse.

9.34 ST ABB'S HEAD (m) ✮

North of St Abb's

○ Visited on ....

White tower (9 metres)
Established: 1862
Designed by: David and Thomas Stevenson
Admiralty number: A2850
Light characteristic: Fl W 10s
Grid reference: NT 915 693
Access inside tower: No

Access to site: Access to the lighthouse is by a 3km (1.8 mile) walk from the National Trust for Scotland car park. The car park is just off of the B6438 near Northfield. Vehicular access to the lighthouse is only permitted for those staying at the holiday cottage.

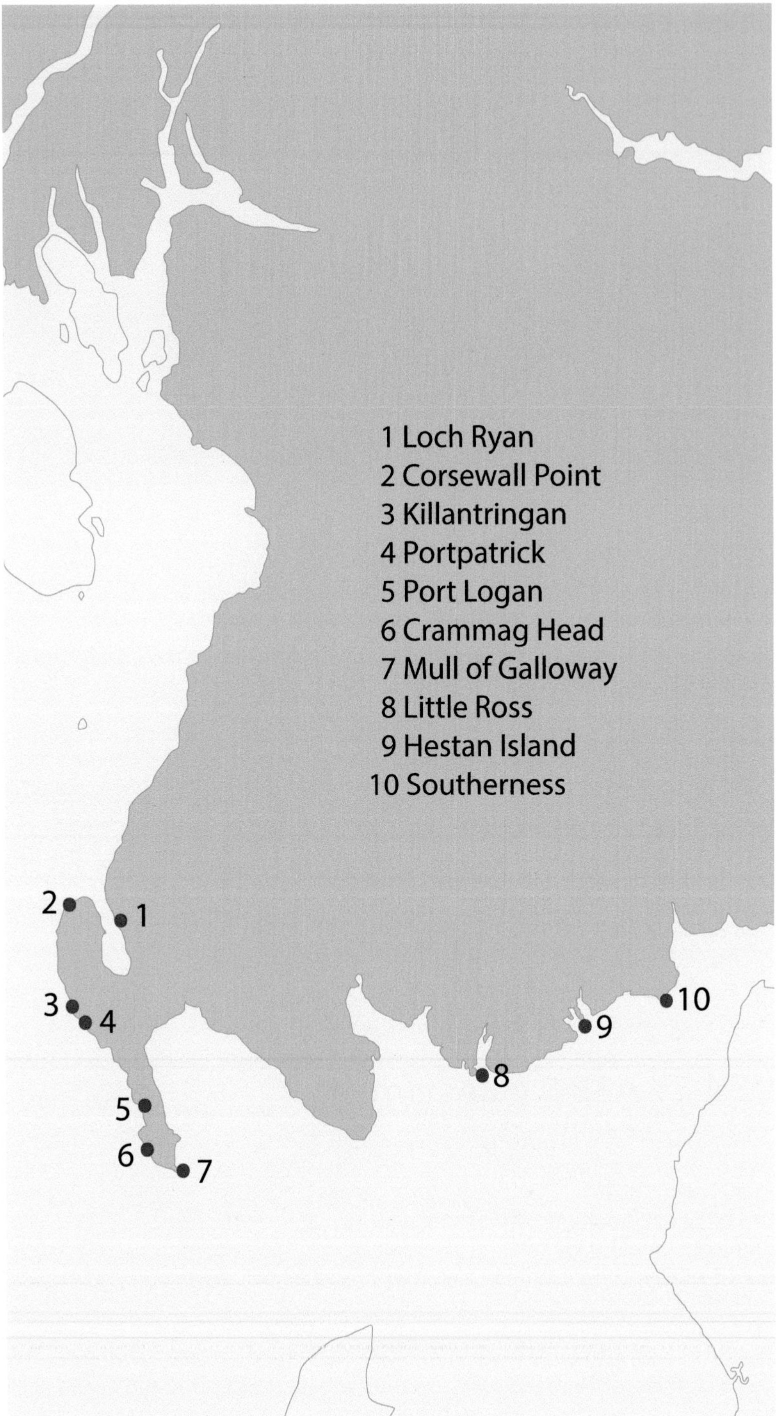
1 Loch Ryan
2 Corsewall Point
3 Killantringan
4 Portpatrick
5 Port Logan
6 Crammag Head
7 Mull of Galloway
8 Little Ross
9 Hestan Island
10 Southerness
2
1
3
4
5
6
7
8
9
10

Region 10

South West Scotland

10.1 LOCH RYAN (CAIRNRYAN POINT) (m) ✯

Near entrance to Cairnryan pier

○ Visited on ....

White tower (15 metres)
Established: 1847
Designed by: Alan Stevenson
Admiralty number: A4592
Light characteristic: Fl(2)R 10s
Grid reference: NX 062 687
Access inside tower: No

Access to site: The lighthouse can be accessed along the shore from the car park a short distance to the north on the A77. It can also be viewed from a slight distance from the road.

10.2 CORSEWALL POINT (m) ✯

North coast of the Rhins of Galloway

○ Visited on ....

White tower (34 metres)
Established: 1817
Designed by: Robert Stevenson
Admiralty number: A4604
Light characteristic: Fl(5)W 30s

Grid reference: NW 981 726
Access inside tower: No

Access to site: The lighthouse is accessible by road, from a northbound turn off of the B738. Parking is available near the lighthouse. The main building, excluding the tower, operates as a hotel and restaurant.

10.3 KILLANTRINGAN (m) ✯

○ Visited on ....

South of Killantringan Bay

Round masonry tower with lantern and gallery (22 metres)
Established: 1900
Designed by: David A and Charles Stevenson
Light discontinued 2007
Grid reference: NW 981 564
Access inside tower: No

Access to site: The lighthouse is accessible from the B738 along a single-track road heading west. Parking is available a short distance from the lighthouse.

10.4 PORTPATRICK (m) ✯

○ Visited on ....

At south entrance to Portpatrick harbour

Round brick tower with lantern and gallery (11 metres)
Established: 1893
Light discontinued 1900
Grid reference: NW 999 539
Access inside tower: No

Access to site: The tower is accessible from South Crescent and parking is available nearby.

10.5 PORT LOGAN (m)

At entrance to Port Logan harbour

○ Visited on ....

Round stone tower with small stone lantern (9 metres)
Established: early 1930s
Light discontinued
Grid reference: NX 095 406
Access inside tower: No

Access to site: The lighthouse is accessible by walking the pier in Port Logan with care. Public parking is available near the entrance to the pier.

10.6 CRAMMAG HEAD (m) ✯

South east coast of the Rhins of Galloway

○ Visited on ....

White tower (7 metres)
Established: 2009
Admiralty number: A4608
Light characteristic: Fl W 10s
Grid reference: NX 089 340
Access inside tower: No

Access to site: The lighthouse is on private land but is visible from single-track roads in the area off of the B7065, south west of Kirkmaiden.

10.7 MULL OF GALLOWAY (m) ✯

Southern Point of the Rhins of Galloway

○ Visited on ....

White round tower (26 metres)
Established: 1830
Designed by: Robert Stevenson

PORT LOGAN

MULL OF GALLOWAY

Admiralty number: A4610
Light characteristic: Fl W 20s
Grid reference: NX 157 304

Access inside tower: The tower is open for tours during the summer months.

Access to site: Access is by road to the south of the B7065. Public parking is available close to lighthouse.

10.8 LITTLE ROSS (i) ✯

○ Visited on ....

On east coast of Little Ross Island, east entrance to Kirkudbright Bay

White tower (20 metres)
Established: 1843
Designed by: Alan Stevenson
Admiralty number: A4634
Light characteristic: Fl W 5s
Grid reference: NX 660 432
Access inside tower: No

Access to site: The island is privately owned. The lighthouse can be seen on a clear day from the single-track road to the south west of Townhead. A closer view would require a chartered boat.

Note: A low light can also be seen on the island, and the former lens from the lighthouse is now on display at the Stewartry Museum in Kircudbright.

10.9 HESTAN ISLAND (t)

South of Almorness Point

◯ Visited on ....

White metal framework tower (7 metres)
Established: 1996
Admiralty number: A4640
Light characteristic: Fl(2)W 10s
Grid reference: NX 839 502
Access inside tower: No

Access to site: Parking is available at the end of the public road towards Almoreness Point, south of Palnackie. The island is tidal allowing only a short period of time for crossing and returning. It is rare for the crossing to be fully cleared of water. From the car park it is a 5km (3 mile) walk to the lighthouse. Distant views of the light on a clear day can be achieved from the tidal Rough Island near Rockcliffe, or from Castle Point to the north east of the island. Castle Point can be reached on foot via a 1.5km (1 mile) walk following the coast to the south from the end of the public road on Rockcliffe sea front where parking is available.

10.10 SOUTHERNESS (t) ✯

South east of Southerness

◯ Visited on ....

White square cylindrical granite tower, black lantern and gallery (10 metres)
Established: 1749
Light discontinued 1936
Grid reference: NX 978 542

Access inside tower: There is occasional public access inside the tower.

Access to site: Public parking is available at the Southerness Holiday Park. From here it is a short walk along the road to the lighthouse.

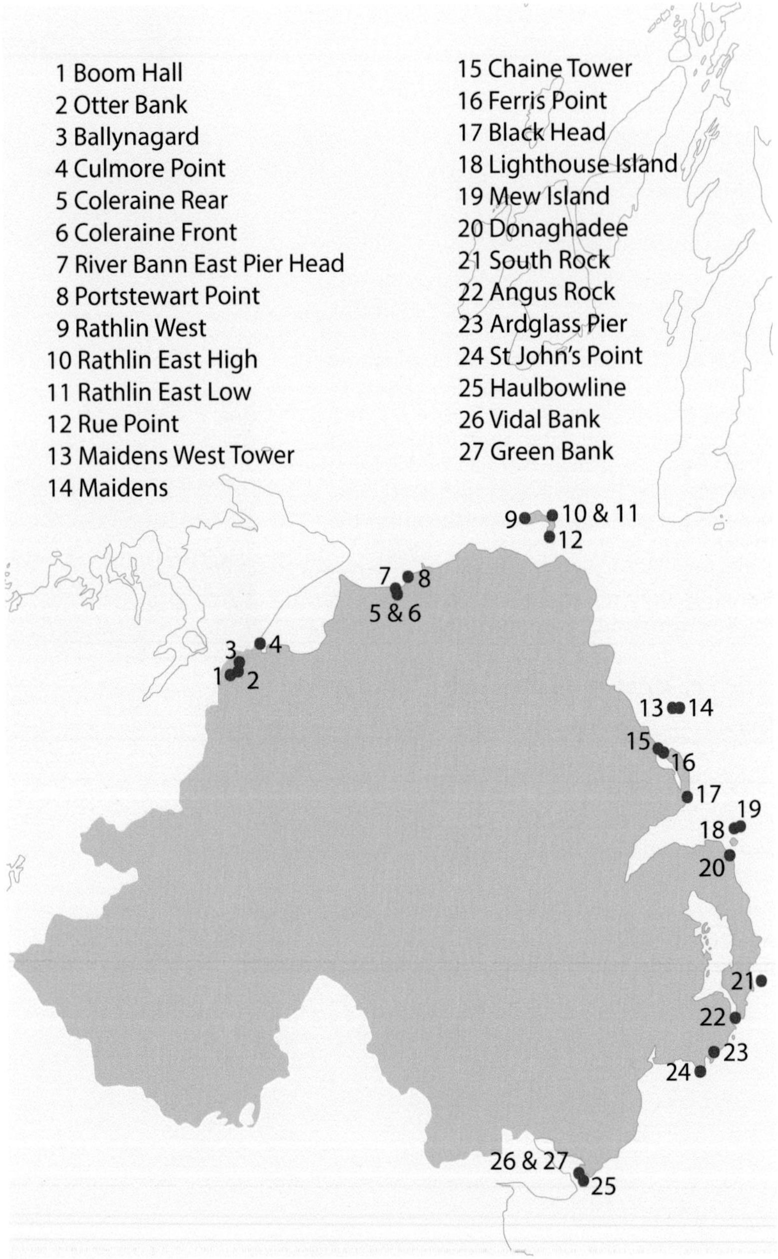
1 Boom Hall
2 Otter Bank
3 Ballynagard
4 Culmore Point
5 Coleraine Rear
6 Coleraine Front
7 River Bann East Pier Head
8 Portstewart Point
9 Rathlin West
10 Rathlin East High
11 Rathlin East Low
12 Rue Point
13 Maidens West Tower
14 Maidens
15 Chaine Tower
16 Ferris Point
17 Black Head
18 Lighthouse Island
19 Mew Island
20 Donaghadee
21 South Rock
22 Angus Rock
23 Ardglass Pier
24 St John's Point
25 Haulbowline
26 Vidal Bank
27 Green Bank
9
10 & 11
12
7
8
5 & 6
3
4
1
2
13
14
15
16
17
18
19
20
21
22
23
24
26 & 27
25

Region 11

Northern Ireland

11.1 BOOM HALL (r)

North west of the Foyle Bridge

White concrete tower, green base
Light discontinued
Grid reference: NV 618 824
Access inside tower: No

◯ Visited on ....

Access to site: The light is best seen from the Foyle Bridge. A closer view would require a chartered boat.

11.2 OTTER BANK (r)

West of Strathfoyle in River Foyle

○ Visited on

....

White concrete tower, red base
Admiralty number: A6138
Light characteristic: Fl R 4s
Grid reference: NV 629 836
Access inside tower: No

Access to site: Best seen by parking at the end of Beechwood Park, Strathfoyle and walking towards the river bank. Be aware that train tracks run alongside the river at this point. A closer view would require a chartered boat. Ballynagard (see 11.3) is also nearby.

11.3 BALLYNAGARD (m)

South of Ballynagard on River Foyle

○ Visited on

....

White concrete tower, green base
Admiralty number: A6136
Light characteristic: Fl W 3s
Grid reference: NV 632 842
Access inside tower: No

Access to site: The lighthouse is on private property. It is best seen from the opposite side of the River Foyle, by parking at the end of Beechwood Park, Strathfoyle and walking towards the river bank. Be aware that train tracks run alongside the river at this point. A closer view would require a chartered boat. Otter Bank (see 11.2) is also nearby.

11.4 CULMORE POINT (m)

South of Culmore on River Foyle

○ Visited on

....

White concrete tower, green base (7 metres)
Established: 1920s
Light discontinued 2012
Grid reference: NV 645 851
Access inside tower: No

Access to site: The light is accessible at the end of Culmore Point Road.

11.5 COLERAINE REAR (RIVER BANN LEADING LIGHTS REAR) (m)

South of entrance to River Bann

◯ Visited on

....

White square concrete tower
Admiralty number: A6074.1
Light characteristic: Oc W 5s
Grid reference: NV 964 960

Access inside tower: Yes. It is possible to step into the ground floor level of the light, although there is little to see.

Access to site: The light is located close to the fifth tee of the Castlerock golf course. The site is accessible either through the Golf Club or by walking along the beach from Castlerock and then alongside the river. After passing the front light (see 11.6) it is a short walk to the south over the dunes to reach the light.

11.6 COLERAINE FRONT (RIVER BANN LEADING LIGHTS FRONT) (m)

South of entrance to River Bann

◯ Visited on

....

White pyramidal metal tower
Admiralty number: A6074
Light characteristic: Oc W 5s
Grid reference: NV 963 961
Access inside tower: No

Access to site: The site is accessible by walking along the beach from Castlerock and then alongside the river. The rear light (see 11.5) is a short walk to the south over the dunes.

11.7 RIVER BANN EAST PIER HEAD (m)

West of Portstewart

◯ Visited on

....

White pyramidal concrete tower (4 metres)
Admiralty number: A6076
Light characteristic: Fl R 5s

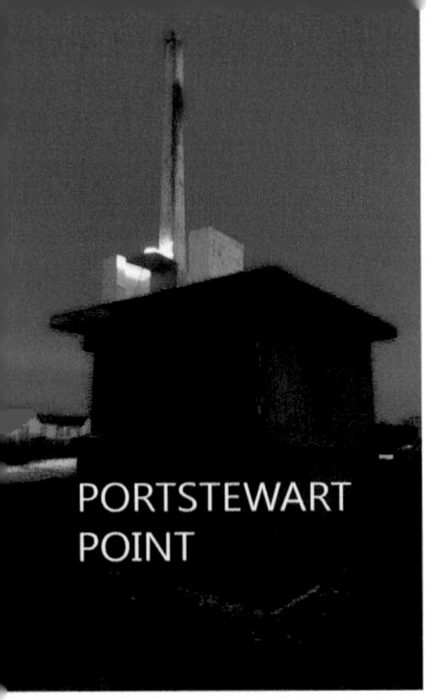

Grid reference: NV 962 968
Access inside tower: No

Access to site: Access is from Portstewart Strand where parking is available. From the car park it is a 3km (1.8 mile) walk to the pier and lighthouse.

11.8 PORTSTEWART POINT (m)

At end of Kinora Terrace, Portstewart

○ Visited on ....

Red square concrete hut (3 metres)
Admiralty number: A6073
Light characteristic: Oc R 10s
Grid reference: NV 997 985
Access inside tower: No

Access to site: The lighthouse is at the end of Kinora Terrace, accessible from the main A2 along Victoria Terrace, and there is public parking next to the light.

11.9 RATHLIN WEST (BULL POINT) (i) ✯

West coast of Rathlin Island

○ Visited on ....

White tower, white lantern at base (18 metres)
Established: 1919
Designed by: Charles William Scott
Admiralty number: A6064
Light characteristic: Fl R 5s
Grid reference: NR 284 091

Access inside tower: Yes. The tower is open to the public as part of the RSPB Seabird Centre during the summer months.

Access to site: Scheduled ferries run from Ballycastle to Rathlin Island throughout the year. Buses run from the ferry terminal to the Seabird Centre during the summer months, or it is a 6.5km (4 mile) walk to the site. Bicycle hire is available on the island if required. Outside of summer there is no access to the site and it is very difficult to see. It is possible to see the lighthouse from certain routes on the Kintyre Express passenger ferry between Ballycastle and Campbeltown, Scotland.

11.10 RATHLIN EAST HIGH (ALTACARRY HEAD HIGH) (i) ✯

North coast of Rathlin Island

◯ Visited on

....

White tower, black band (27 metres)
Established: 1856
Designed by: George Halpin Senior
Admiralty number: A6062
Light characteristic: Fl(4)W 20s
Grid reference: NR 354 088
Access inside tower: No

Access to site: Scheduled ferries run from Ballycastle to Rathlin Island throughout the year. From the ferry terminal it is a 4km (2.5 mile) walk to the lighthouse along a road. Bicycle hire is available on the island if required. The lighthouse compound itself is now private property. The low light (see 11.11) can be seen at the foot of the tower on the seaward side.

11.11 RATHLIN EAST LOW (ALTACARRY HEAD LOW) (i) ✯

North coast of Rathlin Island

◯ Visited on

....

White cast iron lantern (8 metres)
Established: 1856
Designed by: George Halpin Senior
Light discontinued 1894
Grid reference: NR 354 088
Access inside tower: No

Access to site: Scheduled ferries run from Ballycastle to Rathlin Island throughout the year. From the ferry terminal it is a 2.4km (1.5 mile) walk to the lighthouse along a road. Bicycle hire is available on the island if required. The lighthouse compound itself is now private property and the lantern is at the foot of the high light (see 11.10) on the seaward side.

11.12 RUE POINT (RATHLIN SOUTH) (i)

Southern point of Rathlin Island

○ Visited on ....

White eight-sided concrete tower, black bands (11 metres)
Established: 1921
Admiralty number: A6060
Light characteristic: Fl(2)W 5s
Grid reference: NR 338 041
Access inside tower: No

Access to site: Scheduled ferries run from Ballycastle to Rathlin Island throughout the year. From the ferry terminal it is a 4km (2.5 mile) walk to the lighthouse along a road and track. Bicycle hire is available on the island if required.

11.13 MAIDENS WEST TOWER (i)

○ Visited on ....

On west island of the Maidens

Stone tower (18 metres)
Established: 1829
Designed by: George Halpin Senior
Light discontinued 1903
Grid reference: NW 606 659
Access inside lighthouse: No

Access to site: Best seen by chartered boat, although the lighthouse (along with Maidens East Tower, see 11.14) can be seen from a distance on a clear day from the A2 Coast Road north of Larne, or from the Cairnryan to Larne ferry.

11.14 MAIDENS (EAST TOWER) (i) ✮

○ Visited on ....

On east island of the Maidens

White tower, black band (23 metres)
Established: 1829
Designed by: George Halpin Senior
Admiralty number: A6042
Light characteristic: Fl(3)W 15s
Grid reference: NW 613 657
Access inside lighthouse: No

Access to site: By chartered boat only, although the lighthouse (along with Maidens West Tower, see 11.13) can be seen from a distance on a clear day from the A2 Coast Road north of Larne, or from the Cairnryan to Larne ferry.

11.15 CHAINE TOWER (m)

○ Visited on ....

West of entrance to Larne Lough

Grey tower (28 metres)
Established: 1888
Admiralty number: A6031
Light characteristic: Iso WR 5s
Grid reference: NW 563 577
Access inside tower: No

Access to site: The lighthouse is accessible from the corner of Chaine Memorial Road in Larne.

CHAINE TOWER

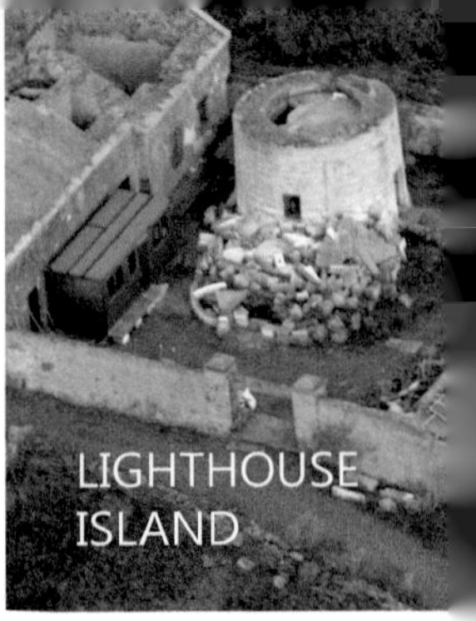

11.16 FERRIS POINT (m)

◯ Visited on ....

East entrance to Larne Lough

White tower, black glass top, white lantern
Established: 1976
Light discontinued 1994
Grid reference: NW 568 573
Access inside tower: No

Access to site: The lighthouse can be seen from Ferris Bay Road. However, ferries approaching Larne offer good views. The lighthouse can be seen from a distance from Chaine Memorial Road in Larne.

11.17 BLACK HEAD (m) ✯

◯ Visited on ....

North of Whitehead

White eight-sided tower (16 metres)
Established: 1902
Designed by: William Douglass
Admiralty number: A6028
Light characteristic: Fl W 3s
Grid reference: NW 628 476
Access inside tower: No

Access to site: It is possible to drive to the site when staying in the holiday cottages at the lighthouse. Otherwise it is a 1km (0.6 mile) walk from Whitehead Car Park, sticking to the inland Blackhead Path.

11.18 LIGHTHOUSE ISLAND (LESSER COPELAND ISLAND) (i)

◯ Visited on ....

On Lighthouse Island within Copeland Islands, north of Donaghadee

Established: 1815
Designed by: George Halpin
Light discontinued 1884
Grid reference: NW 730 391
Access inside tower: No

Access to site: Best seen by chartered boat. The lighthouse can be seen on a clear day from a distance at Orlock Point on the A2, Warren Road, to the east of Groomsport. Mew Island lighthouse (see 11.19) is also visible from here.

11.19 MEW ISLAND (i) ✯

On Mew Island within Copeland Islands, north of Donaghadee

Black tower, white band (37 metres)
Established: 1884
Designed by: William Douglass
Admiralty number: A5976
Light characteristic: Fl(4)W 30s
Grid reference: NW 737 394
Access inside tower: No

○ Visited on ....

Access to site: Best seen by chartered boat. The lighthouse can be seen on a clear day from a distance at Orlock Point on the A2, Warren Road, to the east of Groomsport. The old tower on Lighthouse Island (see 11.18) is also visible from here. The tower can also be seen from ferries to/from Belfast.

Note: The old lens from Mew Island lighthouse is now on display as part of Titanic Belfast on the Titanic Walkway.

11.20 DONAGHADEE (m) ✯

On south pier, Donaghadee

White tower (16 metres)
Established: 1836
Admiralty number: A5974
Light characteristic: Iso WR 4s
Grid reference: NW 723 335
Access inside tower: No

○ Visited on ....

Access to site: Parking is available on the Parade. From here it is a short walk to the end of the pier.

11.21 SOUTH ROCK (KILWARLIN) (r)

North east of Kearney Point

Round granite tower with gallery (18 metres)
Established: 1797
Light discontinued 1877
Grid reference: NW 782 058
Access inside tower: No

○ Visited on ....

Access to site: The lighthouse is visible from a distance from the end of Kearney Road. There is parking on the approach to the end of the road and then it is a short walk to the public footpath along the coast. The light can be seen straight ahead. A closer view would require a chartered boat. It may be possible to see the light from a distance from ferries on some routes between Liverpool and Belfast.

11.22 ANGUS ROCK (i)

In Strangford Narrows

○ Visited on ....

White tower, red top (13 metres)
Established: 1983
Admiralty number: A5963.3
Light characteristic: Fl R 5s
Grid reference: SB 709 985
Access inside tower: No

Access to site: The lighthouse is visible from public car parks to the north and south of Kilclief. A closer view would require a chartered boat.

11.23 ARDGLASS PIER (m) ☆

On north pier, Ardglass

○ Visited on ....

Tower (10 metres)
Established: 1885
Admiralty number: A5962
Light characteristic: Iso WRG 4s
Grid reference: SB 652 913
Access inside tower: No

Access to site: A public car park is near the entrance to the north pier, off of Quay Road. From the car park the entrance to the pier is just to the north.

11.24 ST JOHN'S POINT (m) ☆

South of Killough

○ Visited on ....

Black round tower, yellow bands (40 metres)
Established: 1844
Designed by: George Halpin Senior
Admiralty number: A5958
Light characteristic: Q(2)W 7.5s
Grid reference: SB 616 874
Access inside tower: No

Access to site: The lighthouse compound is private property, but there is parking just outside of the compound. The lighthouse can be reached by

following Rossglass Road South from the A2 at Rossglass. St John's Point is signposted from here.

11.25 HAULBOWLINE (r) ✯

Entrance to Carlingford Lough

○ Visited on

....

Grey granite tower (34 metres)
Established: 1824
Designed by: George Halpin Senior
Admiralty number: A5928
Light characteristic: Fl(3)W 10s
Grid reference: SB 329 659
Access inside tower: No

Access to site: Public parking is available on Ameracam Lane, which provides the best vantage point for seeing the lighthouse. A closer view of the lighthouse would require a chartered boat. At low tide it is possible to walk along the beach to the west to see the Vidal Bank (see 11.26) and Green Island (see 11.27) lights.

11.26 VIDAL BANK (r)

South of Greencastle

○ Visited on

....

Red triangle and white lantern on framework tower (8 metres)
Admiralty number: A5932
Light characteristic: Oc W 3s
Grid reference: SB 322 671
Access inside tower: No

Access to site: The Vidal Bank and Green Island (see 11.27) lights are best seen at low tide by parking on Ameracam Lane in Kilkeel and walking west along the beach.

11.27 GREEN ISLAND (r)

South of Greencastle

○ Visited on

....

Red triangle and white lantern on framework tower (14 metres)
Admiralty number: A5934
Light characteristic: Oc W 3s
Grid reference: SB 319 674
Access inside tower: No

Access to site: The Green Island and Vidal Bank (see 11.26) lights are best seen at low tide by parking on Ameracam Lane in Kilkeel and walking west along the beach.

1 Point of Ayre Low
2 Point of Ayre High
3 Ramsey North Pier
4 Ramsey South Pier
5 Maughold Head
6 Laxey Pier
7 Laxey Breakwater
8 Douglas Battery Pier
9 Douglas Head
10 Derbyhaven Breakwater
11 Langness
12 Castletown Irish Quay
13 Castletown New Pier
14 Port St Mary Inner Pier
15 Port St Mary Alfred Pier
16 Chicken Rock
17 Calf of Man High
18 Calf of Man
19 Calf of Man Low
20 Port Erin Raglan Pier
21 Port Erin Front
22 Peel Breakwater
23 Peel Castle Jetty

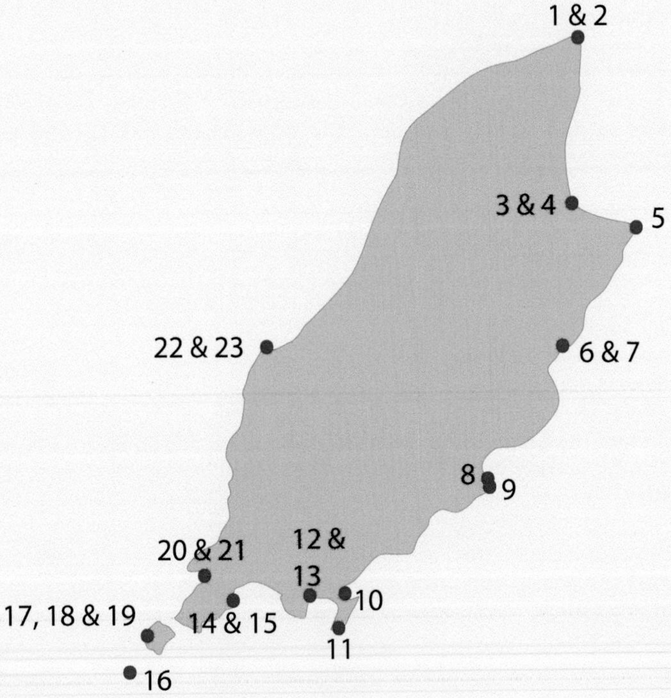

Region 12

Isle Of Man

12.1 POINT OF AYRE LOW (THE WINKIE) (i) ☆

Northern point of Isle of Man

Two-stage octagonal cylindrical tower with lantern and double gallery (10 metres)
Established: 1889
Light discontinued 2010
Grid reference: NX 467 050
Access inside tower: No

○ Visited on ....

Access to site: Access to the lighthouse is along Cranstal Road from Brides. Parking is available a short distance from the lighthouse. Point of Ayre High (see 12.2) is a short distance away.

12.2 POINT OF AYRE HIGH (i) ☆

○ Visited on ....

South west of low light, northern point of Isle of Man

White tower, two red bands (30 metres)
Established: 1819
Designed by: Robert Stevenson
Admiralty number: A4720
Light characteristic: Fl(4)W 20s
Grid reference: NX 464 049
Access inside tower: No

Access to site: Access to the lighthouse is along Cranstal Road from Brides. Parking available a short distance from the lighthouse. The low lighthouse (see 12.1) is a short distance away.

12.3 RAMSEY NORTH PIER (i) ☆

○ Visited on ....

At end of north pier, Ramsey

White tower, black base (8 metres)
Established: 1864
Admiralty number: A4798
Light characteristic: Q G
Grid reference: SC 456 947
Access inside tower: No

Access to site: Parking is available on Mooragh Promenade. From here it is a short walk to the lighthouse. The South Pier lighthouse (see 12.4) is visible and accessible on foot from here.

12.4 RAMSEY SOUTH PIER (i) ☆

○ Visited on ....

At end of south pier, Ramsey

White tower, red band, black base (8 metres)
Established: 1876
Admiralty number: A4796
Light characteristic: Iso R 5s
Grid reference: SC 456 946
Access inside tower: No

Access to site: Parking is available on either Market Place or Queen's Promenade with a short walk to the lighthouse. The North Pier lighthouse (see 12.3) is visible and accessible on foot from here.

12.5 MAUGHOLD HEAD (i) ✯

Eastern point of Isle of Man

○ Visited on ....

White tower (23 metres)
Established: 1914
Designed by: David A and Charles Stevenson
Admiralty number: A4786
Light characteristic: Fl(3)W 30s
Grid reference: SC 498 914
Access inside tower: No

Access to site: The lighthouse is on private land but can be seen from the cliff path. Parking is available near the church in Maughold village, from which a short walk along Lighthouse Road takes you to the lighthouse gate.

12.6 LAXEY PIER (i) ✯

At north end of Laxey Bay

○ Visited on ....

White tower, red band (4 metres)
Established: 1951
Admiralty number: A4784
Light characteristic: Oc R 3s
Grid reference: SC 442 837
Access inside tower: No

Access to site: A public car park is at the entrance to the short pier. The breakwater lighthouse (see 12.7) is visible from here.

12.7 LAXEY BREAKWATER (i) ✯

At north end of Laxey Bay

○ Visited on ....

White tower, green band (4 metres)
Established: 1951
Admiralty number: A4785
Light characteristic: Oc G 3s
Grid reference: SC 444 836
Access inside tower: No

Access to site: The breakwater can be reached on foot from the pier by walking along Tent Road, across the bridge and up the narrow Breeze Hill road. There is a path off of Breeze Hill heading east, which takes you to the breakwater and lighthouse. The pier lighthouse (see 12.6) is visible from here.

12.8 DOUGLAS BATTERY PIER (i) ✯

○ Visited on ....

Halfway along Battery Pier, Douglas harbour

White tower, red band
Established: 1960s
Admiralty number: A4772
Light characteristic: Q R
Grid reference: SC 390 751
Access inside tower: No

Access to site: The pier can be accessed on foot with care from South Quay. The lighthouse can also be seen from ferries arriving at and departing from Douglas terminal. Douglas Head lighthouse (see 12.9) is a short walk away.

12.9 DOUGLAS HEAD (i) ✯

○ Visited on ....

South of Douglas

White tower (20 metres)
Established: 1892
Designed by: David A Stevenson
Admiralty number: A4770
Light characteristic: Fl W 10s
Grid reference: SC 390 748
Access inside tower: No

Access to site: Parking available on Fort Anne Road. A footpath from here takes you to the lighthouse. The Battery Pier lighthouse (see 12.8) is a short walk from here.

12.10 DERBYHAVEN BREAKWATER (i-t) ✯

At south end of breakwater, Derbyhaven

○ Visited on ....

White tower, green band
Established: 1946
Admiralty number: A4764
Light characteristic: Iso G 2s
Grid reference: SC 288 677
Access inside tower: No

Access to site: The breakwater and lighthouse are visible, and accessible at low tide from Airport Road, Derbyhaven.

12.11 LANGNESS (i) ✯

○ Visited on ....

At Dreswick Point, south east of Castletown

White tower (19 metres)
Established: 1880
Designed by: David and Thomas Stevenson
Admiralty number: A4762
Light characteristic: Fl(2)W 30s
Grid reference: SC 282 653
Access inside tower: No

Access to site: The lighthouse can be accessed by following the road south from Derbyhaven. Public parking is available along this road, with a 1km (0.6 mile) onward walk on a tarmac road takes you to the lighthouse.

Notes: Between the car park and lighthouse you will pass the Herring Tower, which was built as a daymark.

The old parabolic mirrors from the lighthouse are now on display at the Museum of Scottish Lighthouses, Fraserburgh.

12.12 CASTLETOWN IRISH QUAY (i) ✯

At end of north pier, Castletown

○ Visited on ....

White tower, red band (4 metres)
Admiralty number: A4761
Light characteristic: Oc R 4s
Grid reference: SC 267 674
Access inside tower: No

Access to site: Short stay (Farrant's Way) and long stay (near the harbour) car parks are signposted from The Parade/Queen Street. From both car parks it is a short walk to both the Irish Quay and New Pier lighthouses (see 12.13).

12.13 CASTLETOWN NEW PIER (i) ✯

○ Visited on ....

At end of south pier, Castletown

White tower, red band (9 metres)
Admiralty number: A4760
Light characteristic: Oc R 15s
Grid reference: SC 267 673

Access inside tower: No

Access to site: Short stay (Farrant's Way) and long stay (near the harbour) car parks are signposted from The Parade/ Queen Street. From both car parks it is a short walk to both the New Pier and Irish Quay lighthouses (see 12.12).

12.14 PORT ST MARY INNER PIER (i) ✯

○ Visited on ....

At end of inner pier, Port St Mary

White tower, red band (8 metres)
Established: 1886
Admiralty number: A4756
Light characteristic: Oc R 3s
Grid reference: SC 211 677
Access inside tower: No

Access to site: The pier is open to the public. Parking is available a short walk from the pier on Lime Street. The Alfred Pier light (see 12.15) is visible from the lighthouse and a short walk away.

12.15 PORT ST MARY ALFRED PIER (i) ✯

○ Visited on ....

At end of Alfred Pier, Port St Mary

White rower, red band (7 metres)
Established: 2018
Light characteristic: Oc R 10s
Grid reference: SC 215 675
Access inside tower: No

Access to site: The pier is accessible from Lime Street. The Inner Pier light (see 12.14) is visible from here and a short walk away.

12.16 CHICKEN ROCK (r) ✯

South west of Caigher Point, Calf of Man

○ Visited on ....

Granite tower (44 metres)
Established: 1875
Designed by: David and Thomas Stevenson
Admiralty number: A4748
Light characteristic: Fl W 5s
Grid reference: SC 143 639
Access inside tower: No

Access to site: The lighthouse can be seen from the Calf of Man lights (see 12.17, 12.18 and 12.19). Boatmen operating trips to the Calf of Man may be willing to sail around to the lighthouse. Landing on the rock requires very calm sea conditions. It may also be possible to see the lighthouse at a distance from ferries on certain routes between Liverpool and Belfast.

12.17 CALF OF MAN HIGH (i) ☆

◯ Visited on ....

North east of Calf of Man light

Round stone tower with lantern and gallery (18 metres)
Established: 1818
Designed by: Robert Stevenson
Light discontinued 1875
Grid reference: SC 150 657
Access inside tower: No

Access to site: Regular passenger only boat trips run from Port Erin or Port St Mary during the summer months. Once on the island it is a 2km (1.25 mile) walk along a path to the lighthouse. The more modern (see 12.18) and low (see 12.19) lighthouses are nearby, and Chicken Rock (see 12.16) can be seen from here.

12.18 CALF OF MAN (i) ☆

◯ Visited on ....

Western point of Calf of Man

White octagonal tower with lantern and gallery on square building (11 metres)
Established: 1968
Designed by: Peter Hyslop
Light discontinued 2007
Grid reference: SC 149 657
Access inside tower: No

Access to site: Regular passenger only boat trips run from Port Erin or Port St Mary during the summer months. Once on the island it is a 2km (1.25 mile) walk along a path to the lighthouse. The old high (see 12.17) and low (see 12.19) lighthouses are nearby, and Chicken Rock (see 12.16) can be seen from here.

12.19 CALF OF MAN LOW (i) ☆

◯ Visited on ....

South east of Calf of Man light

Round stone tower with lantern and gallery (approx. 15 metres)
Established: 1818
Designed by: Robert Stevenson
Light discontinued 1875
Grid reference: SC 149 656
Access inside tower: No

Access to site: Regular passenger only boat trips run from Port Erin and Port St Mary during the summer months. Once on the island it is a 2km (1.25 mile) walk along a path to the lighthouse. The old high (see 12.17) and more modern (see 12.18) lighthouses are nearby, and Chicken Rock (see 12.16) can be seen from here.

12.20 PORT ERIN RAGLAN PIER (i)

○ Visited on ....

On pier at south end of Port Erin ☆

White tower, green band (7 metres)
Established: 1884
Admiralty number: A4744
Light characteristic: Oc G 5s
Grid reference: SC 193 690
Access inside tower: No

Access to site: Access to the pier is on Shore Road where parking is available.

12.21 PORT ERIN FRONT (i) ☆

○ Visited on ....

On beach at Port Erin

White tower, red band (11 metres)
Established: 1884
Admiralty number: A4740
Light characteristic: F R
Grid reference: SC 196 692
Access insider tower: No

Access to site: The lighthouse is on the beach, which can be easily reached from the Promenade.

PORT ERIN FRONT

12.22 PEEL BREAKWATER (i) ✭

○ Visited on

....

At end of breakwater, Peel

White tower (8 metres)
Established: 1896
Admiralty number: A4734
Light characteristic: Oc W 7s
Grid reference: SC 244 847
Access inside tower: No

Access to site: Parking is available a short distance from the lighthouse on or at the entrance to the breakwater. The Peel Castle Jetty light (see 12.23 is also a short distance from here.

12.23 PEEL CASTLE JETTY (i) ✭

○ Visited on

....

On short pier near Peel Castle

White tower, three green bands (4 metres)
Established: 1829
Admiralty number: A4732
Light characteristic: Oc G 7s
Grid reference: SC 243 845
Access inside tower: No

Access to site: Parking is available on or at the entrance to the breakwater. From here it is a short walk to both the Castle Jetty and Breakwater lighthouses (see 12.22).

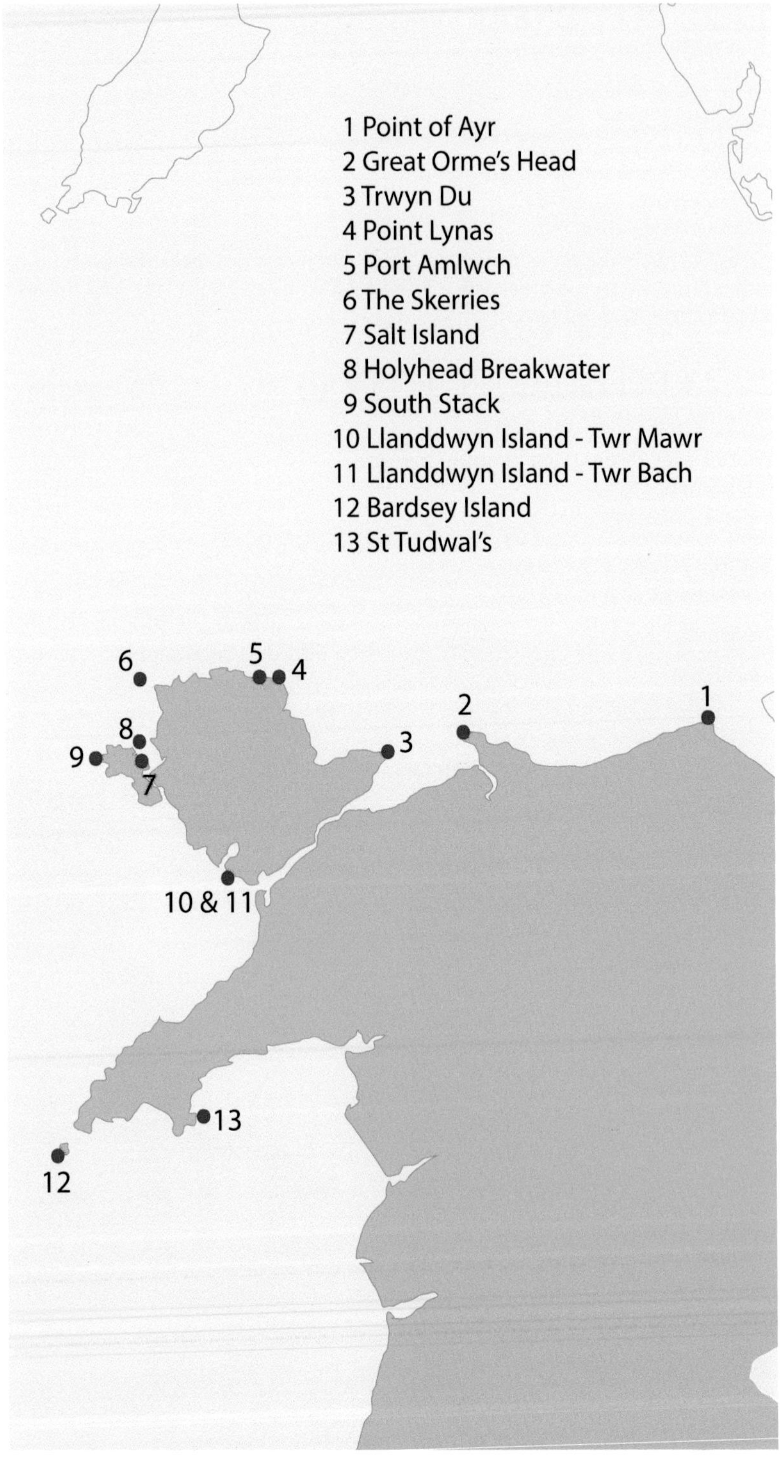
1 Point of Ayr
2 Great Orme's Head
3 Trwyn Du
4 Point Lynas
5 Port Amlwch
6 The Skerries
7 Salt Island
8 Holyhead Breakwater
9 South Stack
10 Llanddwyn Island - Twr Mawr
11 Llanddwyn Island - Twr Bach
12 Bardsey Island
13 St Tudwal's
6
5
4
2
1
8
3
9
7
10 & 11
13
12

Region 13

North Wales

13.1 POINT OF AYR (TALACRE) (t) ☆

North of Talacre

White round tower, red lantern (18 metres)
Established: 1776
Designed by: Henry Turner
Light discontinued 1883
Grid reference: SJ 121 852
Access inside tower: No

○ Visited on ....

Access to site: The lighthouse is accessible on foot over the dunes from Talacre Beach Car Park at the end of Station Road, Talacre.

13.2 GREAT ORME'S HEAD (m) ☆

○ Visited on ....

On Marine Drive, Great Orme's Head

White lantern on side of limestone building (11 metres)
Established: 1862
Designed by: George Fosbery Lyster
Light discontinued 1985
Grid reference: SH 757 844
Access inside tower: The building is now a hotel and access is for customers only.

Access to site: Accessible from the Marine Drive toll road. The site is not open to the public and it is not easy to see the lantern room. The main building is visible from the road.

Note: The old lens from the lighthouse can now be seen at the Visitor Centre at Great Orme.

13.3 TRWYN DU (BLACK POINT, PENMON POINT) (t) ☆

○ Visited on ....

North east of Penmon

White round castellated tower, black bands (29 metres)
Established: 1838
Designed by: James Walker
Admiralty number: A5140
Light characteristic: Fl W 5s
Grid reference: SH 641 815
Access inside tower: No

Access to site: On the approach to Penmon, follow signs for Penmon Point. At Penmon Dovecot you can either park and walk the 1.5km (1 mile) to the coast or pay the toll to drive to the end of the road.

13.4 POINT LYNAS (m) ☆

○ Visited on ....

North west of Llaneilian

White castellated tower (11 metres)
Established: 1835

Designed by: Jesse Hartley
Admiralty number: A5160
Light characteristic: Oc W 10s
Grid reference: SH 479 935
Access inside tower: No, although there is access to the cottages for those staying at the holiday accommodation.

Access to site: From Llaneilian, follow the main road north east and beyond the caravan park. Parking is available just beyond the caravan park. From here it is a 1km (0.6 mile) walk to the lighthouse.

13.5 PORT AMLWCH (m)

○ Visited on ....

End of east pier, Port Amlwch

Square stone tower (11 metres)
Established: 1817
Light discontinued c. 1973
Grid reference: SH 450 935
Access inside tower: No

Access to site: The lighthouse is on the end of the inner pier. Parking is available at the end of Pen Cei beyond the Copper Kingdom Centre. From here it is a short walk back towards the harbour. On a quiet day it may be possible to park near the entrance to the pier.

13.6 THE SKERRIES (i) ☆

○ Visited on ....

North west of Carmel Head, Anglesey

White round tower, red band (23 metres)
Established: 1717
Designed by: William Trench
Admiralty number: A5168
Light characteristic: Fl(2)W 15s
Grid reference: SH 268 948
Access inside tower: No

Access to site: Regular boat trips to see The Skerries operate out of Holyhead. The lighthouse can be seen from a distance from Carmel Head or Holyhead on a clear day. To reach Carmel Head, parking is available at the National Trust car park to the west of Llanfairynghornwy.

13.7 SALT ISLAND (HOLYHEAD ADMIRALTY PIER) (m) ☆

○ Visited on ....

On Salt Island east pier

Round stone tower, green lantern (15 metres)
Established: 1821
Light discontinued
Grid reference: SH 256 829
Access inside tower: No

Access to site: There is no public access to the pier. The lighthouse is best seen from the opposite pier where parking is available on Turkey Shore Road. The lighthouse can also be seen at closer range from the ferry from Holyhead to Dublin.

13.8 HOLYHEAD BREAKWATER (m)

○ Visited on ....

At end of breakwater, Holyhead ☆

White square stone tower, black band (19 metres)
Established: 1873
Admiralty number: A5174
Light characteristic: Fl(3)G 10s
Grid reference: SH 257 848
Access inside tower: No

Access to site: Parking is available at Breakwater Country Park, which is signposted at the west end of Beech Road in Holyhead. From the car park it is a 3.5km (2 mile) walk to the lighthouse along a tarmac road.

13.9 SOUTH STACK (m) ☆

○ Visited on ....

North west of Goferydd

White round tower (28 metres)
Established: 1809
Designed by: Daniel Alexander
Admiralty number: A5204
Light characteristic: Fl W 10s
Grid reference: SH 202 823

Access inside tower: Yes. The lighthouse is open to the public during the summer months and on occasional dates during winter. Guided tours of the lighthouse are available.

Access to site: Parking for the lighthouse is at the end of South Stack Road. From here there are a large number of steps down to the bridge, which can make for a strenuous return journey.

13.10 LLANDDWYN ISLAND - TWR MAWR (t)

South west end of Llanddwyn Island

○ Visited on ....

White round tower (11 metres)
Established: 1845
Light discontinued 1975
Grid reference: SH 385 625
Access inside tower: No

Access to site: Llanddwyn Island is only accessible at low tide. Parking is at Newburgh Warren & Ynys Llanddwyn car park. From here it is a 3km (1.8 mile) walk along the beach or forest road to the island and the lighthouse. The Twr Bach lighthouse (see 13.11) is a little further on.

13.11 LLANDDWYN ISLAND - TWR BACH (t)

South end of Llanddwyn Island

○ Visited on

....

White round tower (12 metres)
Light added: 1975
Admiralty number: A5210
Light characteristic: Fl WR 2.5s
Grid reference: SH 387 624
Access inside tower: No

Access to site: Llanddwyn Island is only accessible at low tide. Parking is at Newburgh Warren & Ynys Llanddwyn car park. From here it is a 3km (1.8 mile) walk along the beach or forest road to the island and the lighthouse. The Twr Mawr lighthouse (see 13.10) is a short distance away.

13.12 BARDSEY ISLAND (i) ✭

On south end of Bardsey Island, south west of Trwyn y Gwyddel

○ Visited on

....

White square tower, red bands (30 metres)
Established: 1821
Designed by: Joseph Nelson
Admiralty number: A5234
Light characteristic: Fl(5)W 15s
Grid reference: SH 111 206
Access inside tower: No

Access to site: Regular boat trips run to Bardsey Island from Porth Meudwy, south west of Aberdaron, during the summer months, but are very weather

LLANDDWYN ISLAND - TWR BACH

BARDSEY ISLAND

dependent. It is advisable to book in advance as the trips are often popular. Parking for Porth Meudwy is at the National Trust Llyn Peninsula car park. From here it is a 1km (0.6 mile) walk to the slipway. Once on the island it is a short grassy walk around the coast to the lighthouse. The compound is private property, but the lighthouse can be viewed from outside.

Note: The old lens from the lighthouse can now be seen at Porth y Swnt Coastal Tourism Centre of Excellence in Aberdaron.

13.13 ST TUDWAL'S (WEST ISLAND) (i) ✯

On St Tudwal's West Island, off of east coast of Bwichtocyn

White round tower (11 metres)
Established: 1877
Admiralty number: A5238
Light characteristic: Fl WR 15s
Grid reference: SH 335 252
Access inside tower: SH 335 252

○ Visited on ....

Access to site: The island is privately owned and landing is not permitted, but boat trips out of Abersoch give a closer view of the island and lighthouse. The lighthouse can also be seen from the mainland on foot from Machroes by following the Wales Coast Path to the south east. The lighthouse should come into view after a short distance, but the closest point is around 2km (1.25 miles) from Machroes.

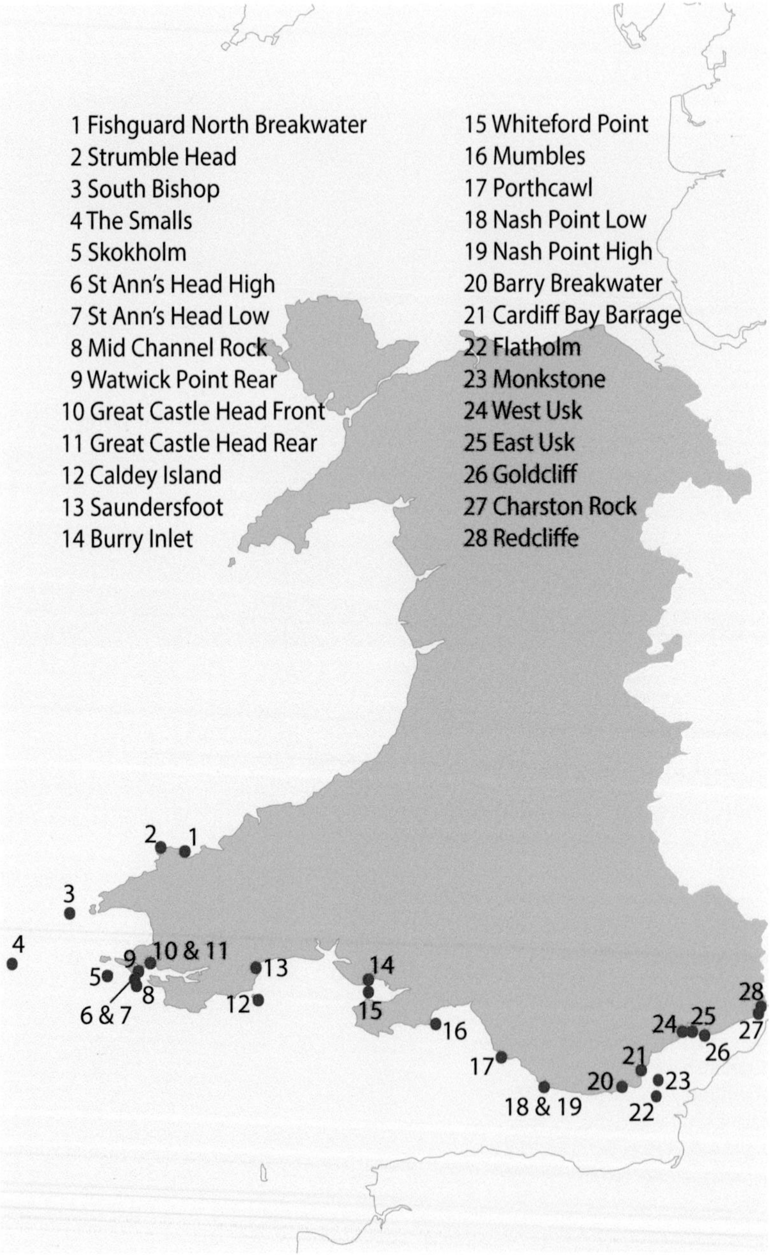
1 Fishguard North Breakwater
2 Strumble Head
3 South Bishop
4 The Smalls
5 Skokholm
6 St Ann's Head High
7 St Ann's Head Low
8 Mid Channel Rock
9 Watwick Point Rear
10 Great Castle Head Front
11 Great Castle Head Rear
12 Caldey Island
13 Saundersfoot
14 Burry Inlet
15 Whiteford Point
16 Mumbles
17 Porthcawl
18 Nash Point Low
19 Nash Point High
20 Barry Breakwater
21 Cardiff Bay Barrage
22 Flatholm
23 Monkstone
24 West Usk
25 East Usk
26 Goldcliff
27 Charston Rock
28 Redcliffe
2
1
3
4
9
10 & 11
5
13
8
6 & 7
12
14
15
16
17
18 & 19
20
21
22
23
24
25
26
27
28

Region 14

South Wales

14.1 FISHGUARD NORTH BREAKWATER (m) ☆

On end of north breakwater, Goodwick

Octagonal concrete tower (20 metres)
Established: 1905
Admiralty number: A5262
Light characteristic: Fl G 4.5s
Grid reference: SM 963 391
Access inside tower: No

○ Visited on ....

Access to site: There is no public access to the breakwater. The lighthouse can be seen from various locations around Fishguard and Goodwick. A good vantage point is Harbour Village, above the port. The ferry between Fishguard and Rosslare (Ireland) passes the lighthouse.

14.2 STRUMBLE HEAD (m) ✯

On the island of Ynys Meicel, north west of Goodwick

◯ Visited on

....

White round tower (17 metres)
Established: 1908
Admiralty number: A5274
Light characteristic: Fl(4)W 15s
Grid reference: SM 893 413
Access inside tower: No

Access to site: There is no access to the bridge that connects the island to the mainland, or to the island on which the lighthouse is located. However, it can be seen clearly from the coast and parking is available at the end of the road from Tresinwen, to the west of Fishguard.

14.3 SOUTH BISHOP (i) ✯

Off of the coast to the south west of St David's

◯ Visited on

....

White round tower (11 metres)
Established: 1839
Designed by: James Walker
Admiralty number: A5276
Light characteristic: Fl W 5s
Grid reference: SM 651 226
Access inside tower: No

Access to site: Best seen by chartered boat. It is also possible to see the lighthouse from the Ramsey Island boat trip, both from the island and from the boat if it sails around Ramsey.

14.4 THE SMALLS (r) ✯

34kms (21 miles) off of the coast at St David's Head

◯ Visited on

....

Grey round tower, red top (41 metres)
Established: 1861
Designed by: James Walker
Admiralty number: A5278
Light characteristic: Fl(3)W 15s
Grid reference: SM 466 089
Access inside tower: No

Access to site: By chartered boat only.

STRUMBLE HEAD

SOUTH BISHOP

THE SMALLS

14.5 SKOKHOLM (i) ✯

○ Visited on ....

Off of the coast, to the south west of Albion Sands

White octagonal tower (18 metres)
Established: 1916
Admiralty number: A5282
Light characteristic: Fl WR 10s
Grid reference: SM 729 046
Access inside tower: For those staying in the holiday accommodation only.

Access to site: The island is only accessible to those who are booked to stay in the holiday accommodation. Otherwise a chartered boat would be required to sail around the island.

14.6 ST ANN'S HEAD HIGH (m)

○ Visited on ....

South of Dale

Black masonry tower with concrete observation box (20 metres)
Established: 1714
Light discontinued 1910
Grid reference: SM 806 029
Access inside tower: For those staying in the holiday accommodation only.

Access to site: The high and low (see 14.7) lighthouses are at the end of the road through St Anne's. Parking is available at the National Trust car park and from here it is a 1.5km (1 mile) walk to both lighthouses.

14.7 ST ANN'S HEAD LOW (m) ✯

○ Visited on ....

South of Dale

White octagonal tower (13 metres)
Established: 1841
Admiralty number: A5284
Light characteristic: Fl WR 5s
Grid reference: SM 807 028
Access inside tower: No

Access to site: The low and high (see 14.6) lighthouses are at the end of road through St Anne's. Parking is available at the National Trust car park and from here it is a 1.5km (1 mile) walk to both lighthouses.

14.8 MID CHANNEL ROCK (r)

○ Visited on ....

South east of St Ann's Head

Black round metal tower, aluminium lantern (18 metres)
Admiralty number: A5285
Light characteristic: Fl(3)G 7s
Grid reference: SM 813 018
Access inside tower: No

Access to site: Best seen from the Pembrokeshire Coast path around St Ann's Head and West Blockhouse Point. A closer view would require a chartered boat.

14.9 WATWICK POINT REAR (m)

○ Visited on ....

South east of Dale

Round tapering concrete tower (50 metres)
Admiralty number: A5286.3
Light characteristic: F W
Grid reference: SM 821 044
Access inside tower: No

Access to site: The lighthouse is on the Pembrokeshire Coast Path. It is best reached via a 1.5km (1 mile) walk from Dale Fort Field Centre, which is at the end of the coast road heading south east from Dale. A longer walk of around 3kms (1.8 miles) to the lighthouse is from St Ann's Head (see 14.6 for directions).

14.10 GREAT CASTLE HEAD FRONT (m)

South east of St Ishmael's

○ Visited on ....

White square tower, black vertical stripe (5 metres)
Established: 1870
Admiralty number: A5287
Light characteristic: Oc W 4s and F WRG
Grid reference: SM 847 060
Access inside tower: No

Access to site: Access to the site is from Bicton Road where there is a small parking area near the road to both the Great Castle Head lighthouses (see 14.11 for details of the rear light). From here proceed on foot for 1.5km (1 mile), turning right at the T junction.

14.11 GREAT CASTLE HEAD REAR (LITTLE CASTLE HEAD) (m)

○ Visited on ....

East south east of St Ishmael's

White round concrete tower (26 metres)
Admiralty number: A5287.1
Light characteristic: OC W 8s
Grid reference: SM 853 066
Access inside tower: No

Access to site: Access to the site is from Bicton Road where there is a small parking area near the road to both the Great Castle Head lighthouses (see 14.10 for the front light). From here proceed on foot for 1km (0.6 miles), turning left at the T junction.

14.12 CALDEY ISLAND (i) ✩

○ Visited on ....

On Caldey Island, off of Giltar Point near Penally

White round tower (16 metres)
Established: 1829
Designed by: Joseph Nelson
Admiralty number: A5328
Light characteristic: Fl(3)WR 20s
Grid reference: SS 143 960
Access inside tower: No

Access to site: Regular boat trips to Caldey operate out of Tenby during the summer months. From the boat it is a 1.5km (1 mile) walk on a road to the lighthouse.

14.13 SAUNDERSFOOT (m)

○ Visited on ....

On south pier, Saundersfoot

Stone cupola (4 metres)
Established: 1848
Admiralty number: A5342
Light characteristic: Fl R 5s
Grid reference: SN 138 047
Access inside tower: No

Access to site: The pier is accessible on foot from the harbour public car park.

14.14 BURRY INLET (m) ☆

○ Visited on ....

On west pier in Burry Port

White tower, red top (7 metres)
Established: 1842
Admiralty number: A5356
Light characteristic: Fl W 5s
Grid reference: SN 445 001
Access inside tower: No

Access to site: The pier is accessible on foot. Parking is available just before the Burry Yacht Club at the entrance to the pier. Access to the car park is from Derwydd Road. It is possible to see Whiteford Point lighthouse (see 14.15) from here on a clear day.

14.15 WHITEFORD POINT (WHITFORD POINT) (t) ☆

○ Visited on ....

South of entrance to River Loughor

Round cast iron tower (13 metres)
Established: 1865
Light discontinued 1926
Grid reference: SS 444 973
Access inside tower: No

Access to site: The lighthouse is only accessible at low tide so it is advisable to check tide times before travelling if you are attempting to reach the lighthouse. The lighthouse is also surrounded by soft sand which should be crossed with care. Parking is available close to The Brittania Inn in Llanmadoc. Walk up the hill for a short distance and turn right, following the path between the white cottages. Follow the signs and at the wooden stile the path divides into two on a circular route, both of which will take you to the lighthouse. The walk is around 5kms (3 miles) from Llanmadoc to the lighthouse. Alternatively, the lighthouse can be seen at a distance from Burry Inlet lighthouse (see 14.14)

14.16 MUMBLES (t) ✯

On end of rocks off of east coast at The Mumbles

○ Visited on

....

White octagonal tower (17 metres)
Established: 1794
Designed by: William Jernegan
Admiralty number: A5358
Light characteristic: Fl(4)W 20s
Grid reference: SS 635 872
Access inside tower: No

Access to site: The lighthouse is visible from the car park on Mumbles Road and a walk across sections of rock and the beach takes you to the lighthouse. It is advisable to check tide times before crossing to the lighthouse and allow plenty of time for returning from the island.

14.17 PORTHCAWL (m) ✯

At end of breakwater, Porthcawl

○ Visited on

....

White 6-sided tower, black base (9 metres)
Established: 1860
Admiralty number: A5402
Light characteristic: F WRG
Grid reference: SS 821 763
Access inside tower: No

Access to site: The breakwater is accessible on foot. Parking is available on Esplanade and Eastern Promenade from which it is a short walk beyond the lifeboat station.

14.18 NASH POINT LOW (m)

South of Marcross

○ Visited on

....

White round tower, lantern removed
Established: 1832

Light discontinued 1920s
Grid reference: SS 918 681
Access inside tower: No

Access to site: On the main road through Marcross, take the turning at The Horseshoe Inn. At the end of the road there is car parking available. From here it is a short walk to the lighthouse. Nash Point High lighthouse (see 14.19) is a little further on.

14.19 NASH POINT HIGH (m) ✯

○ Visited on ....

South of Marcross

White round tower (37 metres)
Established: 1832
Designed by: James Walker
Admiralty number: A5406
Light characteristic: Fl(2)WR 15s
Grid reference: SS 922 680
Access inside tower: The lighthouse Visitor Centre is open for guided tours by prior arrangement. Contact details can be found on the Trinity House website, trinityhouse.co.uk.

Access to site: On the main road through Marcross, take the turning at The Horseshoe Inn. At the end of the road there is car parking available. From here it is a short walk to the lighthouse, passing Nash Point Low lighthouse (see 14.18).

14.20 BARRY BREAKWATER (m) ✯

○ Visited on ....

On end of west breakwater, Barry

White round tower (9 metres)
Established: 1890
Admiralty number: A5412
Light characteristic: Fl W 2.5s
Grid reference: ST 126 665
Access inside tower: No

Access to site: There is no public access to the breakwater. The lighthouse can be seen from Redbrink Crescent or from the headland to the west.

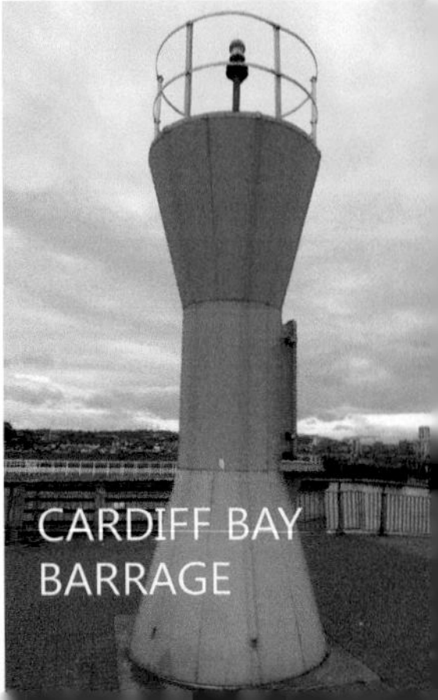

14.21 CARDIFF BAY BARRAGE (m)

○ Visited on ....

On end of north pier, Cardiff Bay Barrage

Green GRP tower
Admiralty number: A5444.31
Light characteristic: Fl G 10s
Grid reference: ST 192 724
Access inside tower: No

Access to site: The pier is accessible on foot. Parking is available at the end of Penarth Portway.

14.22 FLATHOLM (FLAT HOLM) (i) ✯

On Flatholm island, south east of Lavernock Point

○ Visited on ....

White round tower (30 metres)
Established: 1737, re-built 1820
Designed by: William Crispe and Benjamin Lund
Admiralty number: A5426
Light characteristic: Fl(3)WR 10s
Grid reference: ST 222 647
Access inside tower: No

Access to site: Regular boat trips run to the island from both Cardiff and Weston-super-Mare. The closest view of the light from the mainland is at Lavernock Point. However, the lighthouse and island can be seen on a clear day from a number of points along the coast in the area.

14.23 MONKSTONE (r)

○ Visited on ....

East of Penarth

Red GRP column on red masonry tower (25 metres)
Established: 1839
Admiralty number: A5428
Light characteristic: Fl W 5s
Grid reference: ST 236 690
Access inside tower: No

Access to site: The lighthouse is visible from a distance from Penarth. A closer view would require a chartered boat.

14.24 WEST USK (m) ✯

◯ Visited on ....

West of entrance to River Usk

White tapering brick tower
Established: 1821
Designed by: James Walker
Light discontinued 1922
Grid reference: ST 311 829
Access inside tower: For those staying in the lighthouse accommodation only.

Access to site: The access road to West Usk lighthouse is private. The lighthouse can be seen from outside the grounds by walking from the south west. Parking is available off of Beach Road to the south of St Brides. From here it is a 2km (1.2 mile) walk along a path to reach the lighthouse.

14.25 EAST USK (m) ✯

◯ Visited on ....

East of entrance to River Usk

White round tower (13 metres)
Established: 1893
Admiralty number: A5454
Light characteristic: Fl(2)WRG 10s
Grid reference: ST 331 828
Access inside tower: No

Access to site: The lighthouse is within the grounds of RSPB Newport Wetlands, which is accessible from West Nash Road. From the Centre, follow the paths to the south east to reach the lighthouse.

14.26 GOLDCLIFF (m)

◯ Visited on ....

South of Goldcliff

White square tower
Established: circa 1924
Light discontinued

Grid reference: ST 372 819
Access inside tower: No

Access to site: Access to the site is on foot from Seawall. From Goldcliff, head east along Goldcliff Road. At the sharp left corner turn right towards Seawall. At the end of the road there is parking available. From here it is a short walk along a road and then grass to reach the lighthouse.

14.27 CHARSTON ROCK (BLACK ROCK) (r)

East of Portskewett

◯ Visited on ....

White round stone tower, black stripe (7 metres)
Established: 1868
Admiralty number: A5535
Light characteristic: Fl W 3s
Grid reference: ST 520 881
Access inside tower: No

Access to site: The lighthouse can be seen from just beyond the car park at the end of Black Rock Road in Portskewett.

14.28 REDCLIFFE (m) ✯

North east of Portskewett

◯ Visited on ....

Black metal framework tower, white daymarks and lantern (10 metres)
Established: 1910
Admiralty number: A5536
Light characteristic: F Bu
Grid reference: ST 523 896
Access inside tower: No

Access to site: The light is best viewed on foot from the car park at the end of Black Rock Road in Portskewett. Nearby is Charston Rock lighthouse (see 14.27). From here follow the coastal path north for around 1.6km (1 mile). The light is visible from the point at which the path meets the railway line.

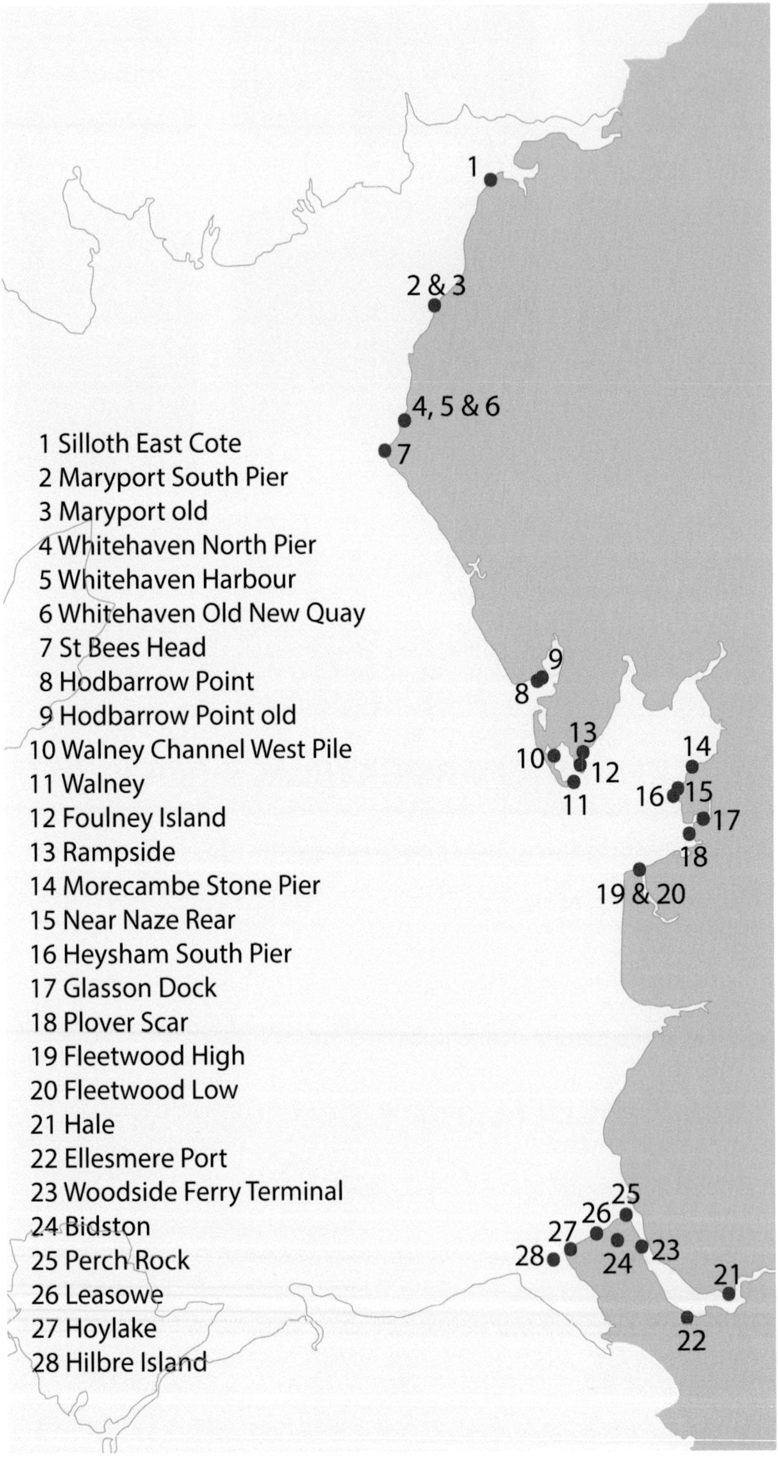
1 Silloth East Cote
2 Maryport South Pier
3 Maryport old
4 Whitehaven North Pier
5 Whitehaven Harbour
6 Whitehaven Old New Quay
7 St Bees Head
8 Hodbarrow Point
9 Hodbarrow Point old
10 Walney Channel West Pile
11 Walney
12 Foulney Island
13 Rampside
14 Morecambe Stone Pier
15 Near Naze Rear
16 Heysham South Pier
17 Glasson Dock
18 Plover Scar
19 Fleetwood High
20 Fleetwood Low
21 Hale
22 Ellesmere Port
23 Woodside Ferry Terminal
24 Bidston
25 Perch Rock
26 Leasowe
27 Hoylake
28 Hilbre Island
1
2 & 3
4, 5 & 6
7
8
9
10
11
12
13
14
15
16
17
18
19 & 20
21
22
23
24
25
26
27
28

Region 15

North West England

White structure on piles (12 metres)
Established: 1997
Admiralty number: A4671
Light characteristic: F G
Grid reference: NY 115 548
Access inside tower: No

○ Visited on

....

Access to site: The lighthouse is on the green just next to Skinburness Road. Car parking is available a short distance to the south of the lighthouse.

15.2 MARYPORT SOUTH PIER (m)

○ Visited on

....

At end of south pier, Maryport

Metal framework tower
Established: 1996
Admiralty number: A4676
Light characteristic: Fl W 1.2s
Grid reference: NY 028 369
Access inside tower: No

Access to site: The pier is accessible with parking close to the lighthouse on Salmoor Way. The old lighthouse (see 15.3) is very close.

15.3 MARYPORT OLD (m) ✩

○ Visited on

....

Near entrance to south pier, Maryport

Octagonal cylindrical cast iron tower with lantern, mounted on octagonal stone base (11 metres)
Established: 1846
Light discontinued
Grid reference: NY 030 368
Access inside tower: No

Access to site: There is parking very close to the lighthouse on Salmoor Way. The south pier lighthouse (see 15.2) is a short walk from here.

15.4 WHITEHAVEN NORTH PIER (m)

○ Visited on

....

At end of the north pier, Whitehaven

White round tower (6 metres)
Established: 1841
Admiralty number: A4700
Light characteristic: Fl R 5s
Grid reference: NX 969 187
Access inside tower: No

Access to site: The pier is accessible on foot. It is a short walk from the Queen's Dock car park on North Shore Road. From the lighthouse you can see the Whitehaven Harbour (see 15.5) and the Old New Quay (see 15.6) lighthouses are visible from the lighthouse.

15.5 WHITEHAVEN HARBOUR (m)

○ Visited on

....

At end of west pier, Whitehaven ✯

White round tower (14 metres)
Established: 1841
Admiralty number: A4698
Light characteristic: Fl G 5s
Grid reference: NX 967 187
Access inside tower: No

Access to site: The pier is accessible on foot. The Wellington Inn car park on West Strand is close to the pier entrance. The Old New Quay lighthouse (see 15.6) is close to the entrance to the pier, and the North Pier lighthouse (see 15.4) is visible from the lighthouse.

15.6 WHITEHAVEN OLD NEW QUAY (m)

On Old New Quay, Whitehaven

○ Visited on

....

Round stone tower with gallery
Established: 1742
Light discontinued
Grid reference: NX 968 185
Access inside tower: No
Access to site: No

Access to site: There is no access to the short pier. However, the lighthouse can be seen clearly from the north pier. The Wellington Inn car park on West Strand is close. The Whitehaven Harbour lighthouse (see 15.5) is a short walk along the west pier from here, and the North Pier lighthouse (see 15.4) is visible from the lighthouse.

Note: The old watch tower, often mistaken for a lighthouse, can also be seen from here.

15.7 ST BEES HEAD (m) ✯

○ Visited on

....

North west of St Bees

White round tower (17 metres)
Established: 1866
Designed by: Joseph Nelson

WHITEHAVEN OLD NEW QUAY

ST BEES HEAD

Admiralty number: A4710
Light characteristic: Fl(2)W 20s
Grid reference: NX 942 144
Access inside tower: No

Access to site: The approach roads to the lighthouse are privately owned, but there is an honesty box car park at Tarnflat Hall Farm, a short distance to the east of the lighthouse. Access is westbound off of Main Street in Sandwith. Alternative access is on foot along the cliff top path from St Bees. This walk is around 4kms (2.5 miles) from the car park on Beach Road in St Bees to the lighthouse.

15.8 HODBARROW POINT (m) ✯

○ Visited on ....

Along Outer Barrier Sea Wall, Haverigg

White and red cast iron tower, gallery and lantern (9 metres)
Established: 1905
Admiralty number: A4805
Light characteristic: Fl W 4s
Grid reference: SD 174 780
Access inside tower: No

Access to site: The lighthouse can be accessed by car or on foot from Haverigg. From the entrance to Port Haverigg Holiday Village, follow the road to the east and then south east around the lake. You will reach the lighthouse after around 2kms (1.25 miles). There is parking available near the entrance to Port Haverigg Holiday Village if you choose to walk to the lighthouse. Alternative access is from Hodbarrow Point Reserve, via the old lighthouse (see 15.9), which is a short walk from this light.

15.9 HODBARROW POINT OLD (m)

○ Visited on ....

South east of Haverigg

Circular stone tower (10.5 metres)
Established: 1866
Light discontinued 1905
Grid reference: SD 180 784
Access inside tower: No

Access to site: The lighthouse is within the Hodbarrow Point Reserve, between Millom and Haverigg. From the car park, follow the main track south east. The tower will be on your left after around 1km (0.6 miles). Alternative access is from Haverigg, via the more modern light (see 15.8), which is a short walk from this light.

15.10 WALNEY CHANNEL WEST PILE (r)

North east of Biggar, Walney Island

◯ Visited on ....

White GRP hut on wooden pile
Admiralty number: A4833.2
Light characteristic: Q R
Grid reference: SD 200 665
Access inside tower: No

Access to site: The light is difficult to see. On a clear day it may be possible to see the light from Mawflat Lane, south of Biggar on Walney, or from Roa Island Road.

15.11 WALNEY (m) ✯

South eastern point of Walney Island

◯ Visited on ....

White stone tower (24 metres)
Established: 1804
Designed by: E Dawson
Admiralty number: A4820
Light characteristic: Fl W 15s
Grid reference: SD 230 621
Access inside tower: No

Access to site: The lighthouse is on private property. However, it can be seen from nearby. Parking is available at South Walney Nature Reserve. From here it is a 2km (1.25 mile) walk east to the closest accessible point to the lighthouse.

15.12 FOULNEY ISLAND (WALNEY CHANNEL MIDDLE RANGE FRONT) (t)

South side of Rampside Sands

◯ Visited on ....

White round GRP tower
Admiralty number: A4824
Light characteristic: Q W
Grid reference: SD 240 648
Access inside tower: No

Access to site: The causeway to Foulney island is covered at high tide, which should be taken into consideration before setting out. A car park is on Roa Island Road on the causeway to Roa Island. From the car park it is a 1km (0.6 mile) walk to the lighthouse. The man-made causeway to the island is uneven and in poor repair in places.

15.13 RAMPSIDE (WALNEY CHANNEL MIDDLE RANGE REAR, THE NEEDLE) (t)

On coast at Rampside

○ Visited on ....

Red brick column, white face
Admiralty number: A4824.1
Light characteristic: Iso W 2s
Grid reference: SD 241 662
Access inside tower: No

Access to site: The lighthouse is visible from Roa Island Road and on street parking is available close by. At the highest of tides the light can be cut off, and the tower is best approached with caution at other times.

15.14 MORECAMBE STONE PIER (m) ✯

At end of pier, Morecambe

○ Visited on ....

Stone tower (8 metres)
Established: 1815
Admiralty number: A4850
Light characteristic: Fl W 5s
Grid reference: SD 426 647
Access inside tower: No

Access to site: The pier is accessible on foot. Parking is available on Marine Road Central, a short walk away from the entrance to the pier.

15.15 NEAR NAZE REAR (HEYSHAM REAR) (m)

North of Heysham harbour

○ Visited on ....

Round stone tower
Light discontinued
Grid reference: SD 403 607
Access inside tower: No

Access to site: The lighthouse is off of Shore Road, via a turn off just before the end of the public road. The best access is on foot from Smithy Lane where a car park is available. The Near Naze Front lighthouse, which no longer has internal access, is a short distance away.

MORECAMBE STONE PIER

NEAR NAZE

15.16 HEYSHAM SOUTH PIER (m) ✯

At end of south breakwater, Heysham

○ Visited on ....

White metal tower, red base (6 metres)
Established: 1904
Admiralty number: A4863
Light characteristic: Oc G 7.5s
Grid reference: SD 396 600
Access inside tower: No

Access to site: There is no public access to the site or the general area. The best view from land is from the end of the public road beyond the Near Naze Rear light (see 15.15). A better view of the lighthouse is from the Heysham to Isle of Man ferry.

15.17 GLASSON DOCK (m)

North east entrance to Glasson Dock

○ Visited on ....

White stone building, white lantern
Established: 1836
Light discontinued
Grid reference: SD 445 563
Access inside tower: No

Access to site: Accessible on foot from the end of Victoria Terrace. The lighthouse is located to the left of the large corrugated metal buildings. Consideration should be given to those working in the area beyond the public road, and care should be taken in this industrial area.

15.18 PLOVER SCAR (t) ✯

At south entrance to River Lune

○ Visited on ....

White stone tower, black lantern (8 metres)
Established: 1847
Admiralty number: A4876
Light characteristic: Fl W 2s
Grid reference: SD 422 543
Access inside tower: No

Access to site: The best access to the lighthouse is by parking just before the entrance to Cockerham Sands Caravan Park, to the north west of Cockerham.

From here it is a 1.8km (1 mile) walk west and then northbound along the coastal path. At low tide it is possible to walk to the lighthouse with care. The lighthouse is also visible from a distance from the end of Slack Lane, beyond Cockerham Sands Country Park.

15.19 FLEETWOOD HIGH (FLEETWOOD REAR, PHAROS) (m) ✯

On Pharos Street, Fleetwood

○ Visited on ....

Stone tower (25 metres)
Established: 1840
Admiralty number: A4892.1
Light characteristic: Iso G 4s
Grid reference: SD 340 483
Access inside tower: No

Access to site: The lighthouse is at the side of Pharos Street. The low light (see 15.20) is visible from here and only a short walk away.

15.20 FLEETWOOD LOW (FLEETWOOD FRONT) (m) ✯

On northern point of Fleetwood Esplanade

○ Visited on ....

Stone tower (12 metres)
Established: 1840
Admiralty number: A4892
Light characteristic: Iso G 2s
Grid reference: SD 339 486
Access inside tower: No

Access to site: There is public access to the Esplanade. The high light (see 15.19) is visible from here and only a short walk away.

15.21 HALE (HALE HEAD) (m) ✯

South of Hale

○ Visited on ....

White tower (17.5 metres)
Established: 1906

Light discontinued 1958
Grid reference: SJ 472 809
Access inside tower: No

Access to site: The lighthouse is at the end of Lighthouse Road in Hale. Lighthouse Road is private but is accessible on foot from the end of Church Road where there is a small parking area. A gate on your right at the end of Lighthouse Road takes you to the coastal path where there are more views of the lighthouse.

Note: The old lens from Hale lighthouse is now at the Merseyside Maritime Museum.

15.22 ELLESMERE PORT (m)

Within National Waterways Museum ☆

Octagonal brick tower (11 metres)
Established: 1880
Light discontinued 1894 ○ Visited on
Grid reference: SJ 405 774 ....
Access inside tower: No

Access to site: The lighthouse is on private property at the end of Lower Mersey Road. It is best seen from the coastal path off of South Pier Road, access is around the side of The Gallery restaurant. There is a public car park on South Pier Road, opposite the National Waterways Museum.

15.23 WOODSIDE FERRY TERMINAL (m) ☆

At Birkenhead ferry terminal. Lantern and tower originally from separate structures

White round tower, dark red lantern
Established: 1986
Light discontinued ○ Visited on
Grid reference: SJ 330 892 ....
Access inside tower: No

Access to site: The lighthouse is behind the ferry terminal. There is parking available in front of the terminal building. The lighthouse is also on the Wirral Circular Trail.

15.24 BIDSTON (m) ○ Visited on

On Bidston Hill ☆ ....

Stone tower
Established: 1873
Designed by: George Fosbery Lyster
Light discontinued 1913
Grid reference: SJ 287 899

WOODSIDE FERRY TERMINAL

BIDSTON

Access inside tower: The lighthouse is open for guided tours every Saturday afternoon from Easter to September, as well as on special occasions. Private tours can also be arranged.

Access to site: The lighthouse is at the end of Wilding Way, off of Boundary Road in Bidston.

15.25 PERCH ROCK (NEW BRIGHTON) (t) ✯

Off of north coast at New Brighton

○ Visited on ....

White round granite tower, red lantern (27 metres)
Established: 1830
Designed by: John Foster
Light discontinued 1973
Grid reference: SJ 309 947
Access inside tower: No

Access to site: The lighthouse is accessible at low tide on New Brighton beach. Parking is available off of Marine Promenade.

15.26 LEASOWE (m) ✯

On Lingham Lane, Leasowe

○ Visited on ....

White round tower (33.5 metres)
Established: 1763
Light discontinued 1908
Grid reference: SJ 253 913
Access inside tower: The tower is open to the public two Sundays each month.

PERCH ROCK

LEASOWE

Access to site: The lighthouse is on Lingham Lane with public parking close to the lighthouse.

15.27 HOYLAKE (m) ✯

On Valentia Road, Hoylake

○ Visited on ....

Octagonal brick tower (16 metres)
Established: 1865
Light discontinued 1886
Grid reference: SJ 214 891
Access inside tower: No

Access to site: The lighthouse is now privately owned but can be viewed from Valentia Road.

15.28 HILBRE ISLAND (t)

South west of Hoylake

○ Visited on ....

White metal framework tower (3 metres)
Admiralty number: A5115
Light characteristic: Fl R 3s
Grid reference: SJ 184 881
Access inside tower: No

Access to site: The island is only accessible at low tide so it is advisable to check tide times before attempting to reach the lighthouse. It is best to allow 3.5 hours and return at least 3 hours before high tide. The walk to the island begins from Marine Lake in West Kirby. Parking is available on South Parade.

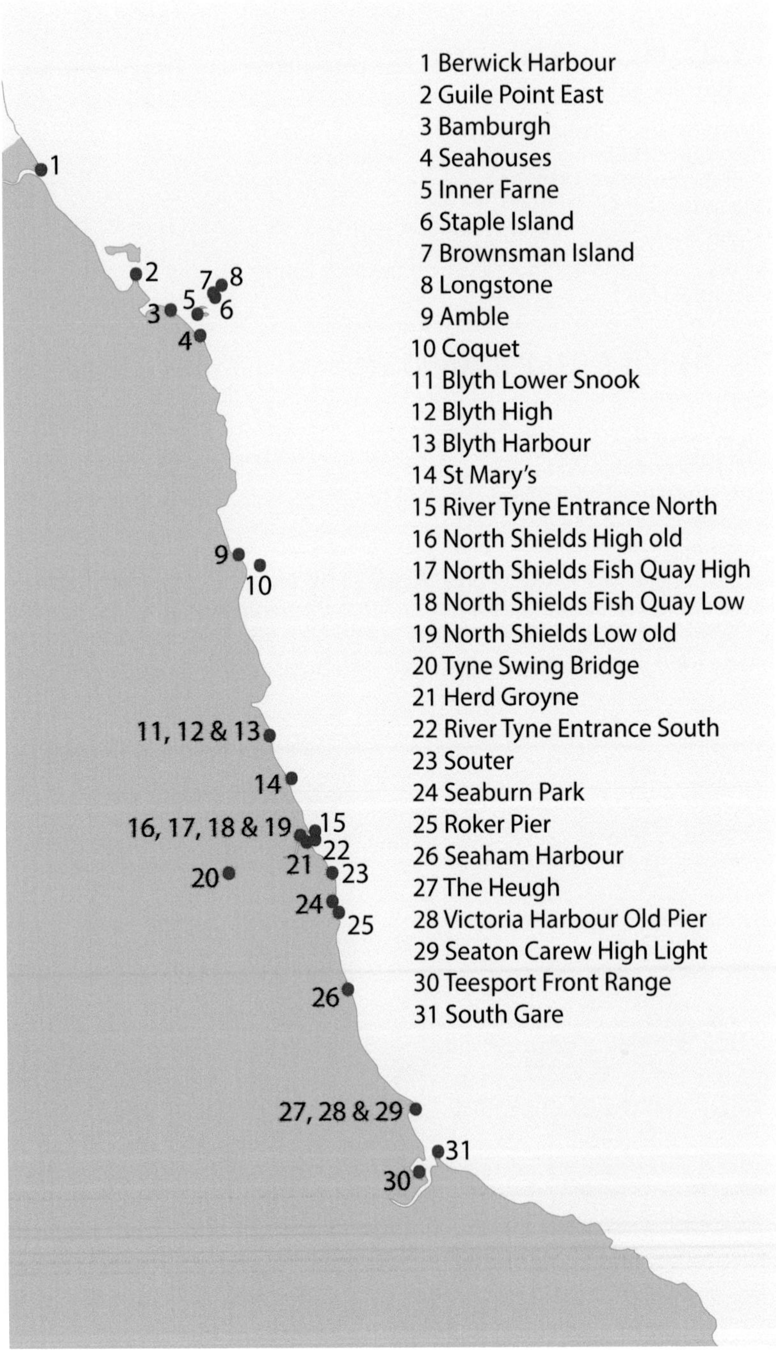
1
2
7
8
3
5
6
4
9
10
11, 12 & 13
14
16, 17, 18 & 19
15
22
21
20
23
24
25
26
27, 28 & 29
31
30
1 Berwick Harbour
2 Guile Point East
3 Bamburgh
4 Seahouses
5 Inner Farne
6 Staple Island
7 Brownsman Island
8 Longstone
9 Amble
10 Coquet
11 Blyth Lower Snook
12 Blyth High
13 Blyth Harbour
14 St Mary's
15 River Tyne Entrance North
16 North Shields High old
17 North Shields Fish Quay High
18 North Shields Fish Quay Low
19 North Shields Low old
20 Tyne Swing Bridge
21 Herd Groyne
22 River Tyne Entrance South
23 Souter
24 Seaburn Park
25 Roker Pier
26 Seaham Harbour
27 The Heugh
28 Victoria Harbour Old Pier
29 Seaton Carew High Light
30 Teesport Front Range
31 South Gare

Region 16

Northumberland And Durham

16.1 BERWICK HARBOUR (m)

At end of breakwater, Berwick

White round stone tower, red cupola and base (13 metres)
Established: 1826
Admiralty number: A2818
Light characteristic: Fl W 5s
Grid reference: NU 011 524
Access inside tower: No

○ Visited on ....

Access to site: The lighthouse is accessible on foot by walking along the breakwater. Parking is available at the entrance to the breakwater.

16.2 GUILE POINT EAST (OLD LAW) (t)

North east of Elwick

○ Visited on ....

Pyramidal stone obelisk (21 metres)
Established: tower: 1826, light: c.1992
Admiralty number: A2816
Light characteristic: Oc RWG 4s
Grid reference: NU 131 405
Access inside tower: No

Access to site: The lighthouse, and its neighbouring obelisk which bears no light, can be accessed at low tide on foot from Ross. The walk is approximately 4kms (2.5 miles) along a clearly marked path and then north along the beach. The lighthouse is the easterly of the two. The two structures can also be seen from the south side of the Holy Island of Lindisfarne.

16.3 BAMBURGH (m) ☆

○ Visited on ....

North of Bamburgh

White building (9 metres)
Established: 1910
Admiralty number: A2810
Light characteristic: Oc(2)WRG 8s
Grid reference: NU 175 359
Access inside tower: No

Access to site: The lighthouse is just off of The Wynding. Parking is available a short distance to the south east of the lighthouse. From here it is possible to see the Farne Islands on a clear day.

16.4 SEAHOUSES (m)

○ Visited on ....

End of north west pier, Seahouses

White tower (8 metres)
Established: 1900
Admiralty number: A2804
Light characteristic: F G

Grid reference: NU 222 323
Access inside tower: No

Access to site: There is public access to the pier with parking a short walk away.

16.5 INNER FARNE (i) ✯

South western of Inner Farne Islands

◯ Visited on ....

White round tower (13 metres)
Established: 1811
Designed by: Daniel Alexander
Admiralty number: A2812
Light characteristic: Fl(2)WR 15s
Grid reference: NU 218 358
Access inside tower: No

Access to site: Landings on Inner Farne are possible in the afternoons from May to July when conditions allow. Boats operate out of Seahouses.

16.6 STAPLE ISLAND (i)

On Staple Island, Farne Islands

◯ Visited on ....

Square stone tower remains (4 metres)
Established: 1784
Light discontinued
Grid reference: NU 238 375
Access inside tower: As the building is in ruins it is advisable to check access with a warden on the island before entering.

Access to site: Landings on Staple Island are possible in the mornings from May to July when conditions allow. Boats operate out of Seahouses.

16.7 BROWNSMAN ISLAND (i)

On Brownsman Isle, Farne Islands

◯ Visited on ....

Square stone tower remains (12 metres)
Established: 1800

Light discontinued 1811
Grid reference: NU 236 379
Access inside tower: No

Access to site: Public access to the island is prohibited. However, the island and the lighthouse can be seen from Farne Island boat trips operating out of Seahouses.

16.8 LONGSTONE (i) ☆

○ Visited on ....

North Farne Islands

Red tower, white band (26 metres)
Established: 1826
Designed by: Joseph Nelson
Admiralty number: A2814
Light characteristic: Fl W 20s
Grid reference: NU 246 390
Access inside tower: The visitor centre is open daily, weather permitting, during the summer months.

Access to site: Routine crossings operate to Longstone, including landing when conditions allow, from Seahouses pier.

Note: The former lens from the lighthouse can now be seen at Think Tank Birmingham Science Museum.

Note: The lighthouse visitor centre also gives access to the former home of Grace Darling, and the RNLI Grace Darling Museum can be found in Bamburgh, just to the north west of Seahouses.

16.9 AMBLE (m)

○ Visited on ....

On south pier head, north of Amble

White round tower, red bands, white base (8 metres)
Admiralty number: A2784
Light characteristic: Fl R 5s
Grid reference: NU 273 051
Access inside tower: No

Access to site: The pier is accessible on foot from the public car park on Harbour Road, although the very end of the pier is closed off. Coquet lighthouse (see 16.10) is visible from here.

16.10 COQUET (i) ✯

○ Visited on ....

On Coquet Island, east of Amble

White square tower, turreted parapet lower half grey (22 metres)
Established: 1841
Designed by: James Walker
Admiralty number: A2780
Light characteristic: Fl(3)WR 20s
Grid reference: NU 292 045
Access inside tower: No

Access to site: It is not possible to land on the island, but boat trips around the island operate from Amble during the summer months. The lighthouse can be seen from a distance from the coastal road south of Amble.

Note: The old lens from the lighthouse is now on display in the Engine Room at Souter lighthouse (see 16.23).

16.11 BLYTH LOWER SNOOK (m)

○ Visited on ....

North of Blyth harbour

White six-sided tower (4 metres)
Admiralty number: A2770
Light characteristic: F Bu
Grid reference: NZ 320 817
Access inside tower: No

Access to site: There is no public access to the site. The light can be seen from the opposite side of the harbour on Quayside where parking is available. From here it is also a fairly short walk to the Blyth High lighthouse (see 16.12).

16.12 BLYTH HIGH (m)

○ Visited on ....

Near Bath Terrace, Blyth

White round tower (12 metres)
Established: 1788
Light discontinued 1984
Grid reference: NZ 320 814
Access inside tower: No

COQUET

BLYTH HIGH

Access to site: The lighthouse is on a lane off of East Park View, behind Bath Terrace.

16.13 BLYTH HARBOUR (BLYTH EAST PIER) (m)

On end of east pier, Blyth ☆

○ Visited on ....

White tower (14 metres)
Established: 1884
Admiralty number: A2754
Light characteristic: Fl(4)W 10s
Grid reference: NZ 328 803
Access inside tower: No

Access to site: There is no public access to the pier, however the lighthouse can be seen from the south beach and south pier with the end of the south pier being the closest vantage point. To access the south pier, car parking is available on Beachway, a small road off of the B1329 (Links Road) to the south of the town. The lighthouse is visible from the car park, but it is a short walk to the south pier from here.

16.14 ST MARY'S (t) ☆

○ Visited on ....

On St Mary's Island, north of Whitley Bay

White round tower (46 metres)
Established: 1898
Light discontinued 1984
Grid reference: NZ 353 754
Access inside tower: Yes. The lighthouse houses a Visitor Centre with self-guided tours of the lighthouse. The centre is open throughout the year, but it is advisable to check the North Tyneside Council website to check opening times before travelling.

Access to site: The lighthouse is accessible on foot via a tidal causeway, to the north of Whitley Bay. There is a car park just before the causeway.

16.15 RIVER TYNE ENTRANCE NORTH (m) ☆

On end of north pier, North Shields

○ Visited on ....

Grey round masonry tower, white lantern (23 metres)
Established: 1908

Admiralty number: A2700
Light characteristic: Fl(3)W 10s
Grid reference: NZ 383 691
Access inside tower: The tower is open on special occasions only.

Access to site: A public car park is at the end of Pier Road. From here it is a short walk to the pier. The pier is closed to the public in rough sea conditions. The River Tyne Entrance South lighthouse (see 16.22) is visible from here.

16.16 NORTH SHIELDS HIGH OLD (m)

Above Western Quay, North Shields

○ Visited on ....

White square building, grey lantern
Established: 1727
Light discontinued: 1810
Grid reference: NZ 361 684
Access inside tower: No

Access to site: The lighthouse is at the junction of Tyne Street and Beacon Street. The new high light (see 16.17) is a short walk away and the new and old low lights (see 16.18 and 16.19) are visible and a short distance from the lighthouse.

16.17 NORTH SHIELDS FISH QUAY HIGH (m)

Above Western Quay, North Shields

○ Visited on ....

White square tower
Established: 1808
Light discontinued 1882
Grid reference: NZ 360 684
Access inside tower: No

Access to site: The lighthouse is at the side of Tyne Street, opposite Laurel Park. On street parking is available. From here it is a short walk to the old high light (see 16.16) as well as the two low lights (see 16.18 and 16.19) in Fish Quay.

16.18 NORTH SHIELDS FISH QUAY LOW (m)

Fish Quay, North Shields

○ Visited on ....

White square tower
Established: 1808

Light discontinued 1882
Grid reference: NZ 363 684
Access inside tower: No

Access to site: The lighthouse is at the side of the road at Fish Quay. The old low light (see 16.19) is just around the corner and both the new and old high lights (see 16.16 and 16.17) are visible and a short walk from the low light. Parking is available near the old low light.

16.19 NORTH SHIELDS LOW OLD (OLD LOW LIGHT) (m)

○ Visited on ....

Fish Quay, North Shields

Rectangular brick building, black side
Established: 1727
Light discontinued 1810
Grid reference: NZ 363 685
Access inside tower: The building has now been converted into the Old Low Light Heritage Centre.

Access to site: The lighthouse is accessible from Clifford Street. The new low light (see 16.18) is just around the corner and both the new and old high lights (see 16.16 and 16.17) are a short walk from here. Parking is available.

16.20 TYNE SWING BRIDGE (m) ✯

○ Visited on ....

On Swing Bridge in Newcastle

Blue lantern on white octagonal building
Established: 1876
Light discontinued
Grid reference: NZ 252 637
Access inside tower: No

Access to site: The lighthouse can be seen when crossing the river on Bridge Street. Parking is available off of Sandhill to the north of the bridge. The lighthouse can also be seen from the train to the east as it crosses the River Tyne at Newcastle.

16.21 HERD GROYNE (m) ✯

○ Visited on ....

On end of inner pier, south of River Tyne

Red pile structure, red and white lantern (15 metres)
Established: 1882
Admiralty number: A2702.5
Light characteristic: Dir Oc W 10s
Grid reference: NZ 369 683
Access inside tower: No

Access to site: The pier is open to the public and is visible and accessible on foot from River Drive in South Shields.

16.22 RIVER TYNE ENTRANCE SOUTH (m) ✯

On end of south pier, South Shields

○ Visited on ....

Grey round stone tower, red and white lantern (12 metres)
Established: 1895
Admiralty number: A2702
Light characteristic: Oc WRG 10s
Grid reference: NZ 384 687
Access inside tower: No

Access to site: The pier is generally open to the public but may be closed in rough sea conditions. Parking is available at the entrance to the pier, off of Sea Road. The River Tyne Entrance North lighthouse (see 16.15) is visible from here.

SOUTER

SEABURN
PARK

ROKER PIER

16.23 SOUTER (m) ✯

Lizard Point, north of Whitburn

◯ Visited on ....

White tower, red bands and lantern (23 metres)
Established: 1871
Designed by: Sir James Douglass
Light discontinued 1988
Grid reference: NZ 408 642
Access inside tower: Yes. The lighthouse is open to the public but closed during the winter months. Tours of the tower are self-guided.

Access to site: The lighthouse is at the side of the A183 coast road at Whitburn.

16.24 SEABURN PARK (SUNDERLAND SOUTH PIER OLD) (m) ✯

Relocated from the end of the south pier in Sunderland to Seaburn Park

◯ Visited on ....

White cast iron tower (15 metres)
Established: 1856
Light discontinued 1903
Grid reference: NZ 408 598
Access inside tower: No

Access to site: The lighthouse is in Seaburn Park, just off of Whitburn Road. There is parking nearby.

16.25 ROKER PIER (m) ✯

At end of Roker pier, Sunderland

◯ Visited on ....

White round tower, 3 red bands and cupola (23 metres)
Established: 1903
Designed by: Henry Hay Wake
Admiralty number: A2681
Light characteristic: Fl W 5s
Grid reference: NZ 416 588
Access inside tower: Yes. Regular tours of both the pier and lighthouse are available during the summer months and occasional days during the winter.

Access to site: The pier is accessible on foot. Parking is available on Marine Walk, a short walk from the pier.

16.26 SEAHAM HARBOUR (m) ✯

At end of north pier, Seaham

◯ Visited on ....

White metal column, black bands (10 metres)
Established: 1905
Admiralty number: A2674
Light characteristic: Fl G 10s
Grid reference: NZ 437 495
Access inside tower: No

Access to site: There is no public access to the north pier. The lighthouse is best seen from the Marina area.

16.27 THE HEUGH (m) ☆

◯ Visited on ....

At the Heugh Battery, Hartlepool

White metal tower (13 metres)
Established: 1927
Admiralty number: A2663
Light characteristic: Fl(2)W 10s
Grid reference: NZ 533 339
Access inside the lighthouse: No

Access to site: The lighthouse is at the end of Moor Terrace where there is parking at the Heugh Battery Museum. From here it is a short walk around the coast to the south west to the Victoria Harbour Old Pier lighthouse. (See 16.28)

Note: The old lens from the lighthouse is now on display at the Hartlepool Maritime Experience.

16.28 VICTORIA HARBOUR OLD PIER (m) ☆

South side of the headland, Hartlepool

◯ Visited on ....

White wooden framework tower, red bands (12 metres)
Established: 1836
Admiralty number: A2664
Light characteristic: Fl WG 3s
Grid reference: NZ 527 335
Access inside tower: No

Access to site: The pier is accessible from York Place where there is on street parking. From here it is a short walk around the coast to the east to The Heugh lighthouse (see 16.27).

16.29 SEATON CAREW HIGH LIGHT (m)

Relocated from Seaton Carew to Hartlepool Marina

◯ Visited on ....

Sandstone tower
Established: 1839
Light discontinued 1892
Grid reference: NZ 515 332
Access inside tower: No

Access to site: Parking is available just off of The Highlight. From here it is a short walk the lighthouse.

16.30 TEESPORT FRONT RANGE (RIVER TEES RANGE FRONT) (m)

○ Visited on ....

East of Billingham

Red metal framework tower, white bands
Admiralty number: A2627
Light characteristic: F R
Grid reference: NZ 537 253
Access inside tower: No

Access to site: There is no public access to the site. The lighthouse is best seen from the road to the south of Seal Sands, off of Seaton Carew Road.

16.31 SOUTH GARE (m) ☆

○ Visited on ....

North west of Redcar

White round tower (13 metres)
Established: 1884
Admiralty number: A2626
Light characteristic: Fl WR 12s
Grid reference: NZ 558 284
Access inside tower: No

Access to site: The lighthouse can be seen by walking north west along the beach for around 5kms (3 miles) from Redcar. There is parking available on Majuba Road. Alternatively, the lighthouse can be seen beyond the end of Tod Point Road, although it should be noted that this is a private road.

SOUTH GARE

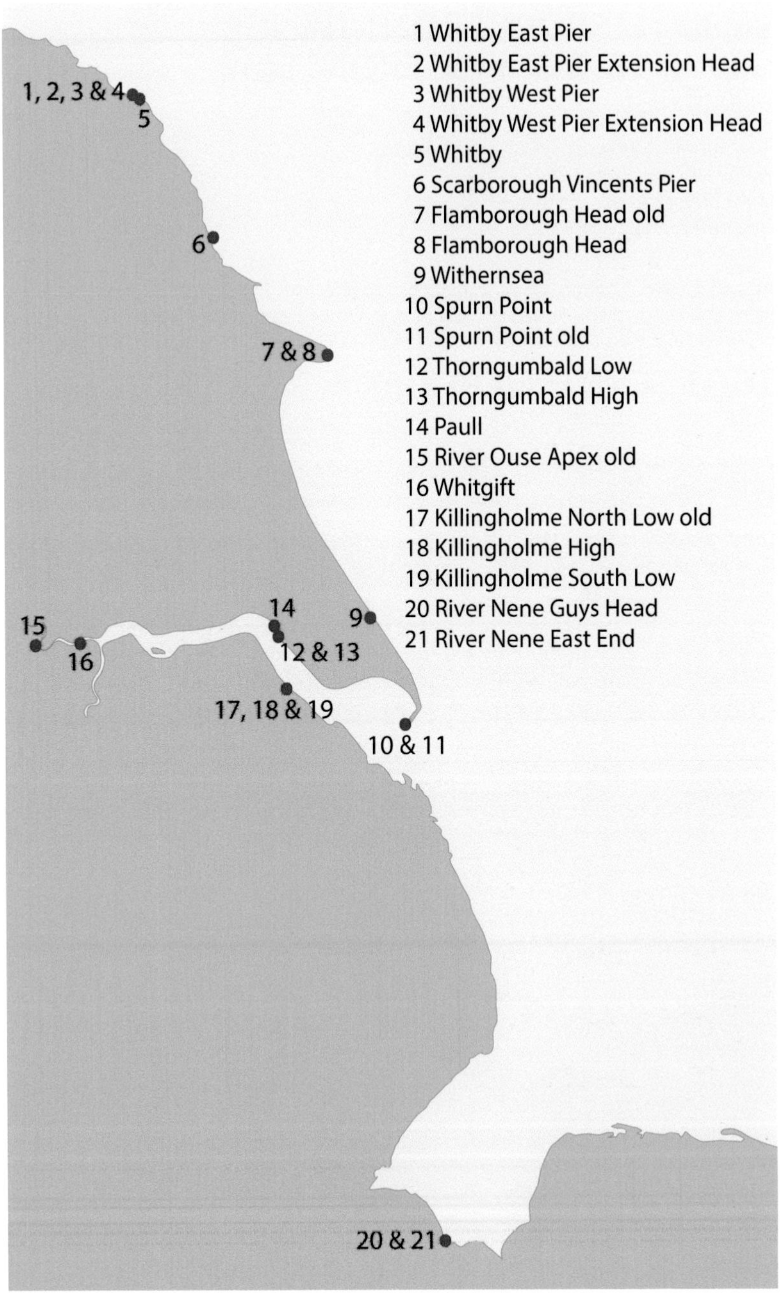
1 Whitby East Pier
2 Whitby East Pier Extension Head
3 Whitby West Pier
4 Whitby West Pier Extension Head
5 Whitby
6 Scarborough Vincents Pier
7 Flamborough Head old
8 Flamborough Head
9 Withernsea
10 Spurn Point
11 Spurn Point old
12 Thorngumbald Low
13 Thorngumbald High
14 Paull
15 River Ouse Apex old
16 Whitgift
17 Killingholme North Low old
18 Killingholme High
19 Killingholme South Low
20 River Nene Guys Head
21 River Nene East End
1, 2, 3 & 4
5
6
7 & 8
9
14
15
16
12 & 13
17, 18 & 19
10 & 11
20 & 21

Region 17

Yorkshire And Lincolnshire

Masonry tower (13 metres)
Established: 1855
Admiralty number: A2598.2
Light characteristic: Fl(2)R 4s
Grid reference: NZ 900 117
Access inside tower: No

○ Visited on
....

Access to site: All four lighthouses in Whitby harbour are visible from Pavillion Drive car park off of North Terrace. This, or Church Street car park, is a good starting point for walking to the piers.

17.2 WHITBY EAST PIER EXTENSION HEAD (i)

In Whitby harbour

◯ Visited on ....

Red wooden tower (7 metres)
Admiralty number: A2598
Light characteristic: Q R
Grid reference: NZ 900 119
Access to tower: No

Access to site: There is no access to the pier extension. The closest viewpoint for seeing the lighthouse is from the end of the west pier extension. All four lighthouses in Whitby harbour are visible from Pavillion Drive car park off of North Terrace. This, or Church Street car park, is a good starting point for walking to the piers.

17.3 WHITBY WEST PIER (m) ✯

◯ Visited on ....

In Whitby harbour

Masonry tower (22 metres)
Established: 1831
Admiralty number: A2599.2
Light characteristic: Fl(2)G 4s
Grid reference: NZ 899 117
Access inside tower: Yes. The lighthouse is open daily, weather permitting, for self-guided tours.

Access to site: All four lighthouses in Whitby harbour are visible from Pavillion Drive car park off of North Terrace. This, or Church Street car park, is a good starting point for walking to the piers.

17.4 WHITBY WEST PIER EXTENSION HEAD (m)

◯ Visited on ....

In Whitby harbour

Green wooden tower (7 metres)
Established: 1914
Admiralty number: A2599

Light characteristic: Q G
Grid reference: NZ 899 119
Access inside tower: No

Access to site: All four lighthouses in Whitby harbour are visible from Pavillion Drive car park off of North Terrace. This, or Church Street car park, is a good starting point for walking to the piers. The end of the west pier extension is the best point for viewing the east pier extension head lighthouse (see 17.2).

17.5 WHITBY (m) ✯

South of Whitby

○ Visited on ....

White octagonal tower (13 metres)
Established: 1858
Designed by: James Walker
Admiralty number: A2596
Light characteristic: Fl WR 5s
Grid reference: NZ 929 101
Access inside tower: No

Access to site: The lighthouse can be accessed along a turn off of Hawsker Lane. The turn-off is recognisable from a stone listing cottage names, including High Light. For a longer walk, the lighthouse can be accessed on foot along the Cleveland Way from Whitby Abbey.

17.6 SCARBOROUGH VINCENTS PIER (m) ✯

Scarborough bay

○ Visited on ....

White round tower (15 metres)
Established: 1931
Admiralty number: A2592
Light characteristic: Iso W 5s
Grid reference: TA 049 886
Access inside tower: No

Access to site: The pier is open to the public. Parking is available on Marine Drive or the West Pier, both a short walk away.

17.7 FLAMBOROUGH HEAD OLD (m)

North east of Bridlington

◯ Visited on ....

Octagonal chalk tower
Established: 1669
Never lit
Grid reference: TA 250 708
Access inside tower: No

Access to site: The lighthouse is within the grounds of Flamborough Head Golf Club. The tower is clearly visible from Lighthouse Road on the approach to the current lighthouse. The modern lighthouse (see 17.8) is a short distance along the road.

17.8 FLAMBOROUGH HEAD (m) ✯

◯ Visited on ....

North east of Bridlington

White round tower (27 metres)
Established: 1806
Designed by: Samuel Wyatt
Admiralty number: A2582
Light characteristic: Fl(4)W 15s
Grid reference: TA 254 707
Access inside tower: The lighthouse is now part of a Trinity House Visitor Centre, including tours of the tower. It is advisable to check opening times on the Trinity House website before travelling.

Access to site: The lighthouse is at the end of Lighthouse Road. There is public parking available at the site. The old lighthouse (see 17.7) is at the side of the road on the approach.

17.9 WITHERNSEA (m) ✯

◯ Visited on ....

In Withernsea town centre

White octagonal tower (39 metres)
Established: 1894
Light discontinued 1972

Grid reference: TA 340 280
Access inside tower: Tours of the lighthouse are available from Easter to October. It is advisable to check opening times before you visit.

Access to site: The lighthouse is at the junction of Hull Street and Arthur Street. There is no parking at the site.

17.10 SPURN POINT (t) ☆

At end of Spurn spit, south of Easington

Black and white round tower (39 metres)
Established: 1895
Designed by: Thomas Matthews
Light discontinued 1985
Grid reference: TA 403 112
Access inside tower: Public tours of the lighthouse are available. It is best to check opening times on the Yorkshire Wildlife Trust website before travelling.

○ Visited on ....

Access to site: Access is at low tide only. It is advisable to check tide times before travelling. Parking is available opposite Spurn Discovery Centre. From here it is a 5km (3.5 mile) walk or cycle to the lighthouse. Alternatively, you can join the Spurn Safari to reach the lighthouse. The old low lighthouse (see 17.11) is on the beach opposite.

17.11 SPURN POINT LOW (t)

At end of Spurn spit, south of Easington

○ Visited on ....

Round stone tower, lantern replaced by water tank (27 metres)
Established: 1852
Light discontinued 1895
Grid reference: TA 402 112
Access inside tower: No

Access to site: Access is at low tide only. It is advisable to check tide times before travelling. Parking is available opposite Spurn Discovery Centre. From here it is a 5km (3 mile) walk or cycle to the lighthouse. Alternatively, you can join the Spurn Safari.

17.12 THORNGUMBALD LOW (m) ✯

West of Thorngumbald

○ Visited on ....

White tower (10 metres)
Established: 1870
Admiralty number: A2506
Light characteristic: Oc W 2s
Grid reference: TA 173 250
Access inside tower: No

Access to site: Parking is available at Paull Holme Strays car park on Thorngumbald Road. From here it is a 1km (0.6 mile) walk to the lighthouse. The walking route to the low light takes you past the high light (see 17.13).

17.13 THORNGUMBALD HIGH (m) ✯

West of Thorngumbald

○ Visited on ....

Red tower on metal framework tower (15 metres)
Established: 1870
Admiralty number: A2508
Light characteristic: Oc W 2s
Grid reference: TA 172 251
Access inside tower: No

Access to site: Parking is available at Paull Holme Strays car park on Thorngumbald Road. From here it is less than 1km (0.6 mile) to the lighthouse. It is a short walk from here to the low lighthouse (see 17.12).

17.14 PAULL (m) ✯

In Paull village

○ Visited on ....

White round brick tower (14 metres)
Established: 1836
Light discontinued 1870
Grid reference: TA 167 262
Access inside tower: No

Access to site: The lighthouse is at the end of Town End Road.

17.15 RIVER OUSE APEX OLD (TRENT FALLS) (m) ✯

○ Visited on ....

Relocated to Yorkshire Riverways Museum, Goole Docks, from the point where the rivers Ouse and Trent meet

Red framework tower (6 metres)
Established: 1933
Light discontinued
Grid reference: SE 742 229
Access inside tower: No

Access to site: The lighthouse is within a closed compound but can be seen from Albert Street on the approach to the Yorkshire Riverways Museum.

17.16 WHITGIFT (m) ✯

○ Visited on ....

North east of Whitgift

Grey masonry and white tower (14 metres)
Admiralty number: A2571.9
Light characteristic: Fl(2)R 4s
Grid reference: SE 821 232
Access inside tower: No

Access to site: Access to the lighthouse is on foot from RSPB Blacktoft Sands car park to the east of Ousefleet. From here it is a 2.5km (1.5 mile) walk westbound to reach the lighthouse. The light can also be seen from the main road in Whitgift.

WHITGIFT

RIVER OUSE APEX OLD

PAULL

17.17 KILLINGHOLME NORTH LOW OLD (m)

East of Killingholme ☆

○ Visited on ....

White round tower (14 metres)
Established: 1836
Light discontinued 1920
Grid reference: TA 178 185
Access inside tower: No

Access to site: The lighthouse is now privately owned, but is visible from Station Road, off of Rosper Road. The high (see 17.18) and south low (see 17.19) lights are nearby.

17.18 KILLINGHOLME HIGH (m) ☆

○ Visited on ....

East of Killingholme

Red tower (24 metres)
Established: 1876
Admiralty number: A2440.1
Light characteristic: Dir WRG
Grid reference: TA 178 182
Access inside tower: No

Access to site: The lighthouse is on private industrial land but is visible from Station Road, off of Rosper Road. The low (see 17.17) and south low (see 17.19) lights are nearby.

17.19 KILLINGHOLME SOUTH LOW (m) ☆

East of Killingholme

○ Visited on ....

White tower (14 metres)
Established: 1836
Admiralty number: A2440
Light characteristic: Iso R 2s
Grid reference: TA 180 181
Access inside tower: No

Access to site: The lighthouse is on private industrial land but is visible from Station Road, off of Rosper Road. The low (see 17.17) and high (see 17.18) lights are nearby.

17.20 RIVER NENE GUYS HEAD (WEST NENE) (m)

In Guy's Head, north of Sutton Bridge

White round tower
Established: 1831
Designed by: John Rennie
Light discontinued
Grid reference: TF 491 258
Access inside tower: No

○ Visited on ....

Access to site: The lighthouse is on West Bank Road in Guy's Head with parking close by. The coastal path from here takes you past the lighthouse. The River Nene East End lighthouse (see 17.21) can be seen from the parking area and the coastal path.

Note: There is debate over whether this tower was ever used as an aid to navigation.

17.21 RIVER NENE EAST END (SIR PETER SCOTT LIGHTHOUSE) (m)

On the east bank of River Nene, north of Sutton Bridge

White round tower
Established: 1831
Designed by: John Rennie
Light discontinued
Grid reference: TF 493 257
Access inside tower: No

○ Visited on ....

Access to site: The Peter Scott Walk begins next to the lighthouse. A parking area is available just before you reach the lighthouse. From here you can also see the Guys Head light (see 17.20).

Note: There is debate over whether this tower was ever used as an aid to navigation.

RIVER NENE GUYS HEAD

RIVER NENE EAST END

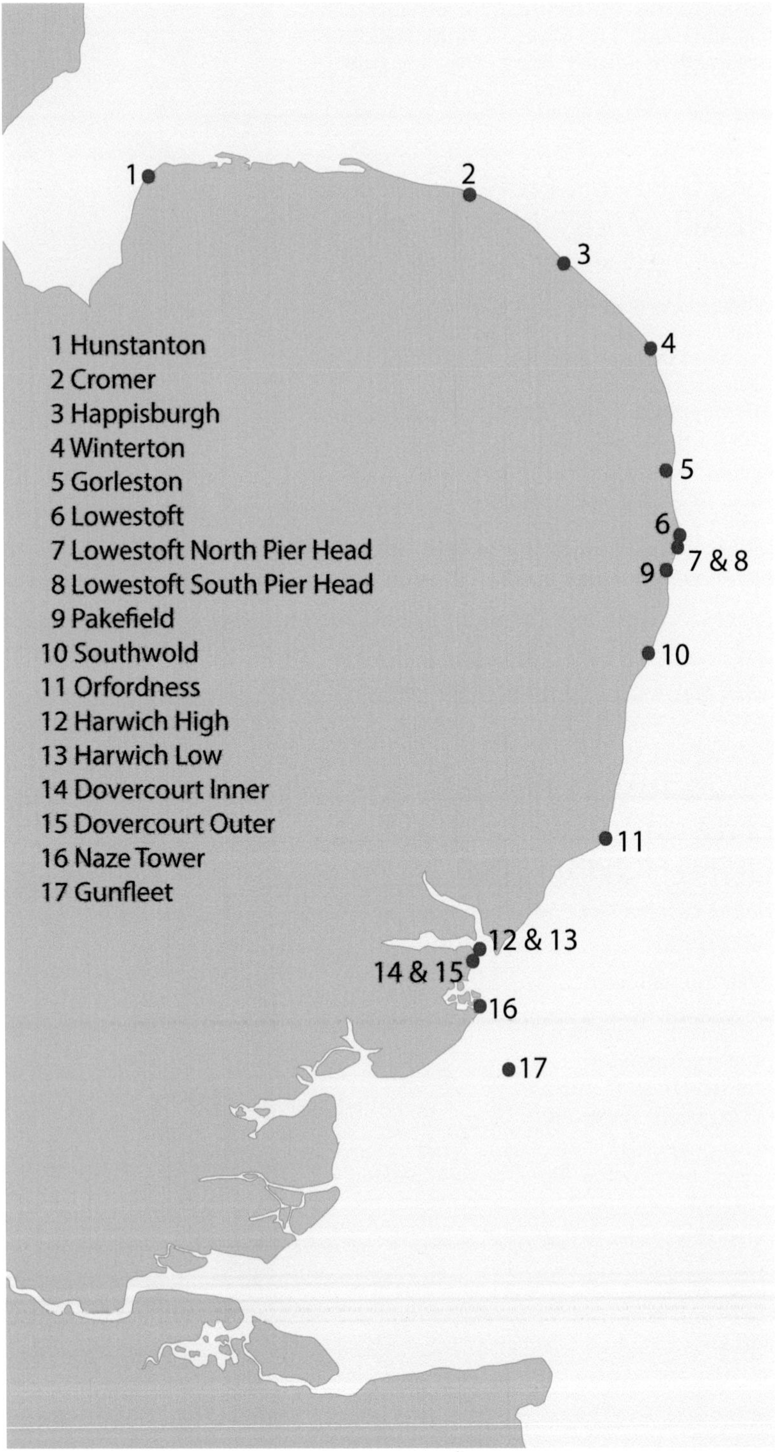
1
2
3
4
5
6
7 & 8
9
10
11
12 & 13
14 & 15
16
17
1 Hunstanton
2 Cromer
3 Happisburgh
4 Winterton
5 Gorleston
6 Lowestoft
7 Lowestoft North Pier Head
8 Lowestoft South Pier Head
9 Pakefield
10 Southwold
11 Orfordness
12 Harwich High
13 Harwich Low
14 Dovercourt Inner
15 Dovercourt Outer
16 Naze Tower
17 Gunfleet

Region 18

East Anglia

18.1 HUNSTANTON (m)

North of Hunstanton

White round tower, lantern removed (19 metres)
Established: 1840
Light discontinued 1921
Grid reference: TF 676 420
Access inside tower: No

○ Visited on ....

Access to site: The lighthouse is at the side of Lighthouse Close. A public car park is available at the end of the road.

18.2 CROMER (m) ✯

South east of Cromer

○ Visited on ....

White octagonal tower (18 metres)
Established: 1833
Admiralty number: A2342
Light characteristic: Fl W 5s
Grid reference: TG 230 416
Access inside tower: No

Access to site: The lighthouse is accessible from Cromer Golf Club. It should be noted that the road from the golf club is private. The lighthouse is visible from the coast path to the south east of the town. This walk begins at the pier, heading south. On street parking is available on the A149 (Runton Road) for the start of the 2km (1.25 mile) walk along the coast to the lighthouse.

18.3 HAPPISBURGH (m) ✯

South of Happisburgh

○ Visited on ....

White tower, 3 red bands (26 metres)
Established: 1791
Designed by: Richard Norris and William Wilkins
Admiralty number: A2336
Light characteristic: Fl(3)W 30s
Grid reference: TG 385 306
Access inside tower: Yes. Organised tours of the lighthouse take place on occasional Sundays and Bank Holidays throughout the summer. Details of opening times can be found on the happisburgh.org website.

Access to site: Access to the lighthouse is on Lighthouse Lane. The lighthouse is clearly visible from the surrounding area. There is a small parking area just outside the lighthouse wall.

Note: Remains of the old low Happisburgh light can been seen at times on the beach to the south east of the lighthouse. A section of the lens from the low light is now on display on the ground floor of Happisburgh lighthouse.

GORLESTON

LOWESTOFT NORTH AND SOUTH PIERS

18.4 WINTERTON (m) ☆

In Winterton-on-Sea

○ Visited on ....

Round brick and masonry tower (21 metres)
Established: 1840
Light discontinued 1921
Grid reference: TG 497 192
Access inside tower: No

Access to site: There is no access to the site. However, the lighthouse can be seen from The Holway or Hillview Drive. Alternatively, is can be viewed from the beach car park to the north of the village, as well as the beach itself.

18.5 GORLESTON (m) ☆

Near mouth of River Yare

○ Visited on ....

Circular red brick tower (21 metres)
Established: 1878
Admiralty number: A2304
Light character: LFl W 2s
Grid reference: TG 530 037
Access inside tower: No

Access to site: The lighthouse is on Quay Road where on street parking is available.

18.6 LOWESTOFT (m) ☆

North of Lowestoft

○ Visited on ....

White tower (16 metres)
Established: 1874
Admiralty number: A2280
Light characteristic: Fl W 15s
Grid reference: TM 551 943
Access inside tower: No

Access to site: The lighthouse is at the side of the A47, Yarmouth Road, but best access for viewing the tower is from Whapload Road to the east of the lighthouse. There is parking here as well as the Lowestoft Maritime Museum. Access to the lighthouse is from Lighthouse Score, a small road to the south of the Maritime Museum. At the end of this road some steps take you to the lighthouse.

18.7 LOWESTOFT NORTH PIER HEAD (m) ☆

At end of north pier, Lowestoft harbour

○ Visited on ....

White tower on small pavilion (9 metres)
Established: 1840s
Admiralty number: A2286
Light characteristic: Oc G 5s
Grid reference: TM 552 927
Access inside tower: No

Access to site: Access to the north pier is prohibited. The lighthouse is best seen from the south pier (see 18.8, Lowestoft South Pier Head, for access details).

18.8 LOWESTOFT SOUTH PIER HEAD (m) ☆

At end of south pier, Lowestoft harbour

○ Visited on ....

White tower on small pavilion (9 metres)
Established: 1840s
Admiralty number: A2284
Light characteristic: Oc R 5s
Grid reference: TM 552 926
Access inside tower: No

Access to site: The majority of the pier is accessible on foot, however the final section and the lighthouse is fenced off, prohibiting access. The lighthouses on both the south and north pier heads can be clearly seen from the fence. Parking is available off of Marine Parade to the south west of the harbour. From here it is a short walk to the south pier.

18.9 PAKEFIELD (m) ☆

In grounds of Pontins Holiday Park

○ Visited on ....

White round tower, blue gallery and lantern, black roof (9 metres)
Established: 1832
Light discontinued 1864
Grid reference: TM 536 887
Access inside tower: It is possible to arrange visits for small groups.

Access to site: The lighthouse is within the grounds of Pontins Holiday Park, which is accessible from London Road. The entrance to the holiday park is often manned, but it is not usually a problem to access the lighthouse (now the coast watch) and the attendant can assist with directions. Once inside the park, at the end of the road, turn right and parking is available at the end of this road. The lighthouse is visible from here.

18.10 SOUTHWOLD (m) ☆

○ Visited on ....

In Southwold town centre

White round tower (31 metres)
Established: 1890
Designed by: Sir James Douglass
Admiralty number: A2272
Light characteristic: Fl(4)WR 20s
Grid reference: TM 510 763
Access inside tower: Yes. The tower is open for guided tours from late March to October. It is advisable to check opening days and times on the Trinity House website before travelling.

Access to site: The lighthouse is on Stradbroke Road. There is no parking at the lighthouse.

18.11 ORFORDNESS (m) ☆

○ Visited on ....

On shingle spit east of Orford

White round tower, red bands (30 metres)
Established: 1792
Designed by: William Wilkins
Light discontinued 2013
Grid reference: TM 450 489
Access inside tower: At time of going to print, occasional organised tours of the lighthouse were taking place. Details of how to book a visit can be found on the website, www.orfordnesslighthouse.co.uk.

Access to site: Regular crossings to Orford Ness on a small passenger boat from Orford operate during the summer months. From the boat it is a 2.5km (1.5 mile) walk to the lighthouse on a good path. The lighthouse can also be viewed from a distance from Orford.

Note: The old lens from the lighthouse is now located at the UN International Maritime Organization Headquarters in London. Their headquarters are open to the public on special occasions.

Note: At time of going to print, the lighthouse is in danger of being lost to the sea.

18.12 HARWICH HIGH (m) ☆

◯ Visited on ....

In Harwich town centre

Nine-sided brick tower, pinnacle roof (27 metres)
Established: 1818
Light discontinued 1863
Grid reference: TM 261 324
Access inside tower: Yes. The light is now home to a Wireless Museum, which offers public tours at weekends during the summer.

Access to site: The lighthouse is at the side of West Street/Main Road. The low light (see 18.13) is a short walk from here.

18.13 HARWICH LOW (m)

◯ Visited on ....

North of entrance to Harwich harbour

Octagonal white tower, black roof (6 metres)
Established: 1818
Light discontinued 1863
Grid reference: TM 263 324
Access inside tower: Yes. The light now houses the Harwich Maritime Museum, which is open to the public at weekends during the summer, and Fridays in July and August.

Access to site: The lighthouse is at the north east end of Marine Parade. The high light (see 18.12) is a short walk from here.

18.14 DOVERCOURT INNER (DOVERCOURT REAR) (m) ☆

◯ Visited on ....

South west of entrance to Harwich harbour

Black six-sided cast iron tower, white wooden lantern, black roof (13 metres)
Established: 1863
Light discontinued 1917
Grid reference: TM 253 309
Access inside tower: No

Access to site: The lighthouse is at the side of Lower Marine Parade in Dovercourt. The outer light (see 18.15) is just to the east.

18.15 DOVERCOURT OUTER (DOVERCOURT FRONT) (t) ☆

○ Visited on ....

South west of entrance to Harwich harbour

Black six-sided cast iron tower, white wooden lantern, black roof (8 metres)
Established: 1863
Light discontinued 1917
Grid reference: TM 255 308
Access inside tower: No

Access to site: The lighthouse is accessible at low tide with care, via the inner lighthouse (see 18.14), which is next to Lower Marine Parade.

18.16 NAZE TOWER (m)

○ Visited on ....

North east of Walton on the Naze

Octagonal red brick tower (26 metres)
Established: 1720
Light discontinued
Grid reference: TM 265 236
Access inside tower: Yes. The tower is open for self-guided tours daily from Easter to October.

Access to site: The lighthouse and car park are accessible by following Naze Park Road and Old Hall Lane from the centre of Walton on the Naze.

Note: There is debate over whether the tower was ever lit.

18.17 GUNFLEET (r)

○ Visited on ....

Six miles east of Frinton on Sea

Screw pile tower beneath six-sided corrugated iron structure, small lantern (22.5 metres)
Established: 1850
Light discontinued 1920
Grid reference: TM 306 131
Access inside tower: No

Access to site: The lighthouse is visible on a very clear day from the Naze Tower (see 18.16) or from Frinton on Sea. A closer view would require a chartered boat.

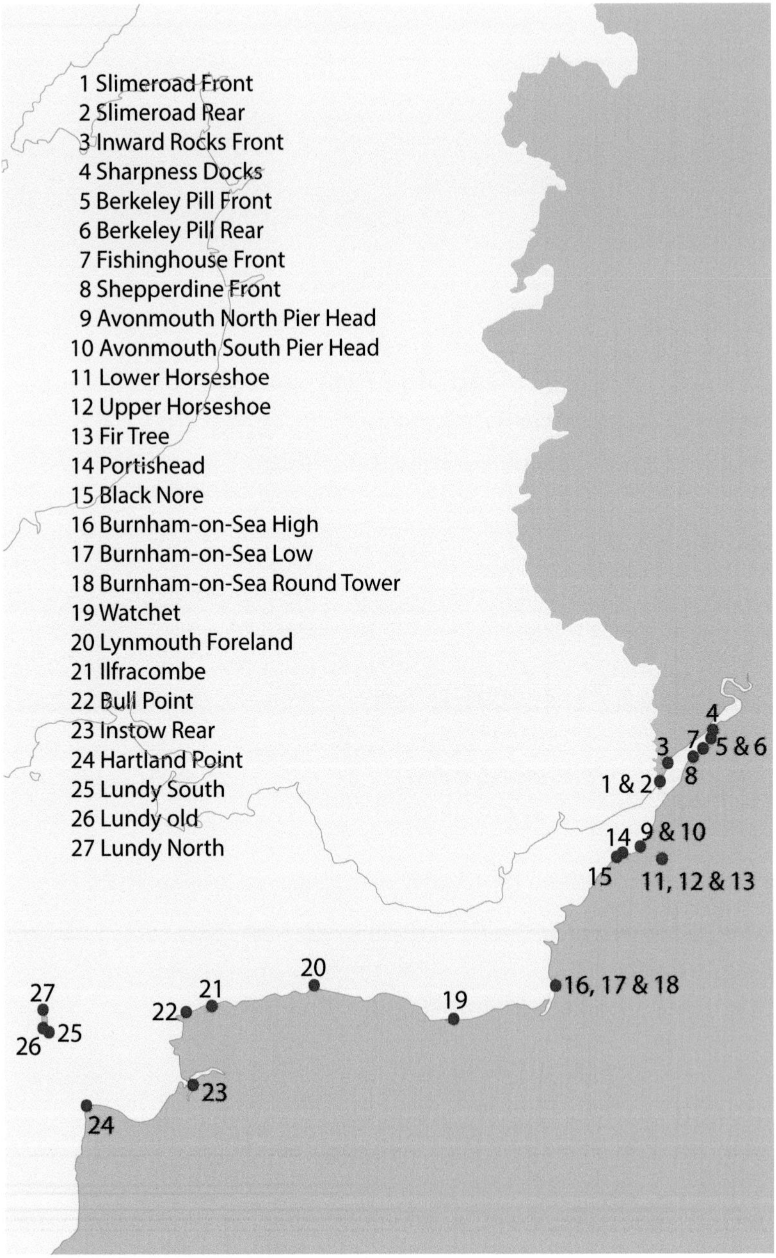
1 Slimeroad Front
2 Slimeroad Rear
3 Inward Rocks Front
4 Sharpness Docks
5 Berkeley Pill Front
6 Berkeley Pill Rear
7 Fishinghouse Front
8 Shepperdine Front
9 Avonmouth North Pier Head
10 Avonmouth South Pier Head
11 Lower Horseshoe
12 Upper Horseshoe
13 Fir Tree
14 Portishead
15 Black Nore
16 Burnham-on-Sea High
17 Burnham-on-Sea Low
18 Burnham-on-Sea Round Tower
19 Watchet
20 Lynmouth Foreland
21 Ilfracombe
22 Bull Point
23 Instow Rear
24 Hartland Point
25 Lundy South
26 Lundy old
27 Lundy North
4
3
7
5 & 6
1 & 2
8
14
9 & 10
15
11, 12 & 13
20
16, 17 & 18
27
21
19
22
26
25
23
24

Region 19

Bristol Channel And North Devon

19.1 SLIMEROAD FRONT (m)

In Beachley

White hut (2 metres)
Established: 1941
Admiralty number: A5547
Light characteristic: F Bu
Grid reference: ST 550 915
Access inside tower: No

○ Visited on ....

Access to site: The light is at the side of Beachley Road opposite the 1st Battalion The Rifles compound. Although it can be seen from the road where the rear light (see 19.2) is also visible, the best view is from the river bank, which is accessible along a small path between the houses on Loop Road. From here it is a 2km (1.25 mile) walk to the light. Care should be taken in the area, with visits avoided where possible at higher tides, due to the light's close proximity to the river.

19.2 SLIMEROAD REAR (m) ✯

◯ Visited on

....

In Beachley

Black metal framework tower, white lantern (8 metres)
Established: 1941
Admiralty number: A4447.1
Light characteristic: F Bu
Grid reference: ST 550 914
Access inside tower: No

Access to site: The lighthouse is within the 1st Battalion The Rifles compound on Beachley Road. There is no access inside the compound, but the light is visible from the road. The top of the front light (see 19.1) can be seen just across the road.

19.3 INWARD ROCKS FRONT (m)

◯ Visited on

....

South east of Tidenham

White round GRP tower (6 metres)
Established: 1985
Admiralty number: A5548
Light characteristic: F W
Grid reference: ST 569 953
Access inside tower: No

Access to site: Access to the lighthouse is via a 1.3km (0.8 mile) walk through fields where there is very likely to be livestock present. The path marked on maps is very unclear on the ground, so the walk is not recommended for those with little or no navigation experience. The start of the walk is through the farm gate opposite the entrance to Tidenham Lane on the A48. The path crosses an electric fence and under the railway line, and the route is not marked. The light can also be seen from a boat along the River Severn.

19.4 SHARPNESS DOCKS (m)

◯ Visited on

....

Relocated close to entrance to Sharpness Tidal Basin

White lantern
Light discontinued

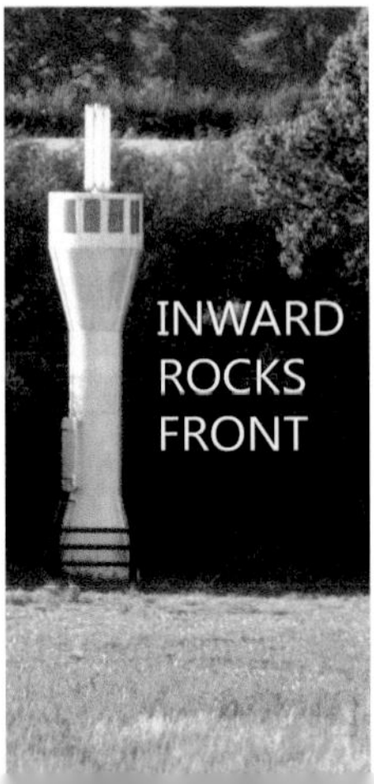

Grid reference: SO 668 021
Access inside tower: No

Access to site: The lantern is just to the north of the Sharpness viewpoint and picnic site parking area, at the end of Great Western Road.

Note: The lantern is likely to have been used on one of the Gloucester Harbour Trustees' lights until it was moved to its current location.

19.5 BERKELEY PILL FRONT (m) ☆

○ Visited on ....

North of Berkeley Power Station

Black metal framework tower, white lantern and daymark (8 metres)
Established: 1937
Admiralty number: A5556
Light characteristic: F G
Grid reference: SO 662 002
Access inside tower: No

Access to site: Both the front and rear (see 19.6) lights are visible from the Severn Way path. The closest starting point for reaching the lights is via the footpath on the north east side of the bridge across Berkeley Pill. The bridge is on the main road out of Berkeley to the west, heading towards the power station. From here it is a 1.5km (1 mile) walk to reach the nearest point to the lights.

19.6 BERKELEY PILL REAR (m) ☆

○ Visited on ....

North of Berkeley Power Station

Black metal framework tower, white lantern (12 metres)
Established: 1937
Admiralty number: A5556.1
Light characteristic: F G
Grid reference: SO 662 001
Access inside tower: No

Access to site: Both the front and rear (see 19.5) lights are visible from the Severn Way path. The closest starting point for reaching the lights is via the footpath on the north east side of the bridge across Berkeley Pill. The bridge is on the main road out of Berkeley to the west, heading towards the power station. From here it is a 1.5km (1 mile) walk to reach the nearest point to the lights.

19.7 FISHINGHOUSE FRONT (m)

○ Visited on ....

West of Berkeley

White GRP cone tower, orange and white daymarks (4 metres)
Established: 1985
Admiralty number: A5552
Light characteristic: F W
Grid reference: ST 644 984
Access inside tower: No

Access to site: The light is on the Severn Way. The closest starting point for reaching the light is along the public footpath near the roundabout as you approach Berkeley Power Station. From here it is a 2.5km (1.5 mile) walk to the light. The rear light, a post with lights at the top is close by.

19.8 SHEPPERDINE FRONT (SHEPERDINE FRONT) (m)

○ Visited on ....

North of Oldbury-on-Severn

Grey enclosure on black round tower (8 metres)
Established: 1996
Admiralty number: A5549
Light characteristic: F W
Grid reference: ST 617 967
Access inside tower: No

Access to site: The light is visible from the Severn Way. The best access is on foot from the end of Shepperdine Road and head north east once you reach the river bank. From here it is a short walk to the light. The rear light, a grey post with lights mounted part way up, is nearby.

19.9 AVONMOUTH NORTH PIER HEAD (m) ☆

At end of pier north of entrance to Avonmouth Docks

○ Visited on ....

Round stone tower (16 metres)
Established: 1908

Admiralty number: A5492
Light characteristic: Fl W 4s
Grid reference: ST 503 791
Access inside tower: No

Access to site: There is no public access to the site or the surrounding area. The light, along with the south pier light (see 19.10) can be seen from a distance on a clear day while travelling eastbound on the M4 but are best seen by boat. Scheduled tourist trips along to the mouth of the river run from Bristol.

19.10 AVONMOUTH SOUTH PIER HEAD (m)

At end of pier south of entrance to Avonmouth Docks ☆

○ Visited on ....

Round stone tower (9 metres)
Established: 1907
Admiralty number: A5492.21
Light characteristic: Oc RG 30s
Grid reference: ST 503 789
Access inside tower: No

Access to site: There is no public access to the site or the surrounding area. The light, along with the north pier light (see 19.9) can be seen from a distance on a clear day while travelling eastbound on the M4 but are best seen by boat. Scheduled tourist trips along to the mouth of the river run from Bristol.

19.11 LOWER HORSESHOE (m)

North bank of River Avon, to the east of Shirehampton

○ Visited on ....

White column (4 metres)
Admiralty number: A5514
Light characteristic: Fl R 2s
Grid reference: ST 539 767
Access inside tower: No

Access to site: The light is best seen by boat. Scheduled tourist trips along the river operate from Bristol. The Upper Horseshoe light (see 19.12) is a little further to the east and the Fir Tree light (see 19.13) is further to the south east. The light can also be seen from the Fir Tree light. All three lights can be seen from the River Avon Trail on the south bank of the river, following the path from Chapel Pill Lane, to the east of Pill.

AVONMOUTH NORTH PIER HEAD

AVONMOUTH SOUTH PIER HEAD

19.12 UPPER HORSESHOE (m)

○ Visited on ....

North bank of River Avon, to the east of Shirehampton

White column (4 metres)
Admiralty number: A5515
Light characteristic: Fl(3)R 6s
Grid reference: ST 542 768
Access inside tower: No

Access to site: The light is best seen by boat. Scheduled tourist trips along the river operate from Bristol. The Lower Horseshoe light (see 19.11) is a little further to the west, and the Fir Tree light (see 19.13) is to the south east. The light can also be seen from the Fir Tree light. All three lights can be seen from the River Avon Trail on the south bank of the river, following the path from Chapel Pill Lane, to the east of Pill.

19.13 FIR TREE (m)

○ Visited on ....

North bank of River Avon, to the west of Sea Mills

White column (4 metres)
Admiralty number: A5516
Light characteristic: Q(2)R 4s
Grid reference: ST 544 766
Access inside tower: No

Access to site: A footpath off of the south side of the A4, Portway, slightly to the north west of the Sylvan Way (A4162) junction. Follow the path and once you have passed under the railway bridge head north west. The light is a short walk from the bridge. The Upper Horseshow (see 19.12) and Lower Horseshow (see 19.11) lights are visible from here. All three lights can be seen from the River Avon Trail on the south bank of the river, following the path from Chapel Pill Lane, to the east of Pill.

19.14 PORTISHEAD (m)

North of Portishead, West of Bristol

Black metal framework tower, white concrete base (9 metres)
Established: 1930
Admiralty number: A5484
Light characteristic: Q(3)W 10s
Grid reference: ST 464 776
Access inside tower: No

◯ Visited on ....

Access to site: Access to the lighthouse is on foot from Esplanade Road where parking is available.

19.15 BLACK NORE (m)

North west of Portishead ✯

White hexagonal cast iron tower on cast iron frame with lantern and gallery (11 metres)
Established: 1894
Light discontinued 2010
Grid reference: ST 446 765
Access inside tower: No

◯ Visited on ....

Access to site: There are a number of access points for the coastal path on which the lighthouse is located. The closest point is through a small gate on Glenwood Rise, close to the junction with Nore Road. For a longer walk, parking is available off of Beach Road West. From here it is a 1.8km (1 mile) walk heading west along the coast.

19.16 BURNHAM-ON-SEA HIGH (m) ✯

North of Burnham-on-Sea

White brick tower, conical roof (30 metres)
Established: 1832
Light discontinued 1993
Grid reference: ST 304 505
Access inside tower: No

◯ Visited on ....

Access to site: The lighthouse is on Berrow Road. It can also be seen from Gore Road and the path from here to the low lighthouse (see 19.17).

BURNHAM-ON-SEA HIGH

19.17 BURNHAM-ON-SEA LOW (m)

○ Visited on ....

On Burnham-on-Sea beach

White square tower, red stripe on piles (9 metres)
Established: 1832
Admiralty number: A5570
Light characteristic: Fl W 7.5s
Grid reference: ST 299 504
Access inside tower: No

Access to site: The lighthouse can be accessed on foot along a small footpath at the end of Gore Road. An alternative path gives access from the corner of Trinity Rise and Pizey Avenue. The high lighthouse (see 19.16) can also be seen from the Gore Road path.

19.18 BURNHAM-ON-SEA ROUND TOWER (m)

○ Visited on ....

On Esplanade, at north end of churchyard in Burnham-on-Sea

White round masonry tower with castellated top
Established: 1801
Light discontinued 1832
Grid reference: ST 304 494
Access inside tower: No

Access to site: The lighthouse is within the grounds of The Round Tower Guest House on the Esplanade where parking is available. The top of the tower can be seen from here, but the best vantage point is from within the grounds of St Andrew's Church.

19.19 WATCHET (m) ✯

○ Visited on ....

At end of Watchet west breakwater

Red six-sided metal tower, white lantern, green cupola (5 metres)
Established: 1862
Admiralty number: A5584
Light characteristic: Oc G 3s
Grid reference: ST 072 436
Access inside tower: No

Access to site: The breakwater is accessible on foot. There is a car park just off of Market Street near the entrance to the breakwater.

19.20 LYNMOUTH FORELAND (FORELAND POINT) (m) ✯

○ Visited on ....

North east of Lynton

White round tower (15 metres)
Established: 1900
Admiralty number: A5590
Light characteristic: Fl(4)W 15s
Grid reference: SS 754 511
Access inside tower: No

Access to site: Car parking is available off of the A39 to the east of Countisbury. From the parking area it is a 2.5km (1.5 mile) walk on a zig zag tarmac road to the lighthouse. The return journey is very hard going. Guests in the lighthouse cottage can drive to the lighthouse.

19.21 ILFRACOMBE (m)

On St Nicholas Chapel, to the west of entrance to Ilfracombe harbour

○ Visited on ……………………

White lantern on chapel (11 metres)
Established: 1819
Admiralty number: A5594
Light characteristic: Fl(2)G 5s
Grid reference: SS 525 479
Access inside tower: The chapel is open to the public, but there is no access to the lantern.

Access to site: The chapel can be seen, and accessed on foot up a steep path, from the car park at the end of The Quay.

19.22 BULL POINT (m) ☆

North of Morthoe

○ Visited on ……………………

White round tower (11 metres)
Established: 1975
Admiralty number: A5600
Light characteristic: Fl(3)W 10s
Grid reference: SS 463 468
Access inside tower: No

Access to site: The lighthouse is accessible on foot from Morthoe, where there is a public car park. From here follow North Morte Road north east and at the end of the road you will see the gate for the lighthouse. From the car park it is a 2.3km (1.4 mile) walk to the lighthouse. The South West Coast Path also passes the lighthouse.

19.23 INSTOW REAR (m)

North end of Instow

○ Visited on ....

White tower (9 metres)
Established: 1964
Admiralty number: A5610.1
Light characteristic: Oc W 10s
Grid reference: SS 478 311
Access inside tower: No

Access to site: The light is at the side of Rectory Lane, just off of Anstey Way.

The Instow Front light is visible from here.

19.24 HARTLAND POINT (m) ☆

North west of Hartland

○ Visited on ....

White brick round tower (17 metres)
Established: 1874
Designed by: Sir James Douglass
Light discontinued 2011
Grid reference: SS 230 278
Access inside tower: No

Access to site: From the village of Titchberry, north of Hartland, follow the road to the north west. At the end of the road there is a car park. It is not possible to access the lighthouse compound, but the lighthouse can be seen from the coastal path in the area.

19.25 LUNDY SOUTH (i) ☆

○ Visited on ....

On south east point

White round tower (16 metres)
Established: 1897
Designed by: Thomas Matthews
Admiralty number: A5618
Light characteristic: Fl W 5s
Grid reference: SS 144 437
Access inside tower: No

Access to site: Lundy island is accessible by passenger ferry three days a week from either Ilfracombe or Bideford from April to October. It is advisable to check the timetable before travelling. Once on the island access to the south lighthouse is via steps just above the pier. It is possible to book accommodation on the island should you wish to visit the island's three lighthouses (see also 19.26 and 19.27) across a number of days.

19.26 LUNDY OLD (i) ☆

○ Visited on ....

On Beacon Hill, Lundy island

Granite tower with gallery and lantern (29 metres)
Established: 1820
Designed by: Daniel Alexander

Light discontinued 1897
Grid reference: SS 132 443
Access inside tower: Yes. The tower is open for self-guided tours.

Access to site: Lundy island is accessible by passenger ferry three days a week from either Ilfracombe or Bideford from April to October. It is advisable to check the timetable before travelling. Once on the island it is a steep walk up a tarmac road and then a short walk on a clearly marked path heading east to the lighthouse. It is possible to book accommodation on the island should you wish to visit the island's three lighthouses (see also 19.25 and 19.27) across a number of days.

19.27 LUNDY NORTH (i) ☆

○ Visited on ....

On northern point of Lundy island

White round tower (17 metres)
Established: 1897
Designed by: Thomas Matthews
Admiralty number: A5616
Light characteristic: Fl W 15s
Grid reference: SS 131 482
Access inside tower: No

Access to site: Lundy island is accessible by passenger ferry three days a week from either Ilfracombe or Bideford from April to October. It is advisable to check the timetable before travelling. Once on the island it is a steep walk up a tarmac road and then on clearly marked tracks heading north. It is a 5.6km (3.5 mile) walk from the ferry to the lighthouse. It is possible to book accommodation on the island should you wish to visit the island's three lighthouses (see also 19.25 and 19.26) across a number of days.

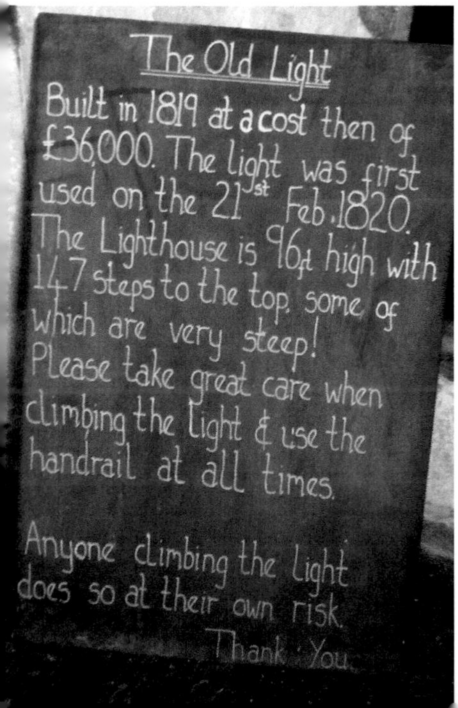

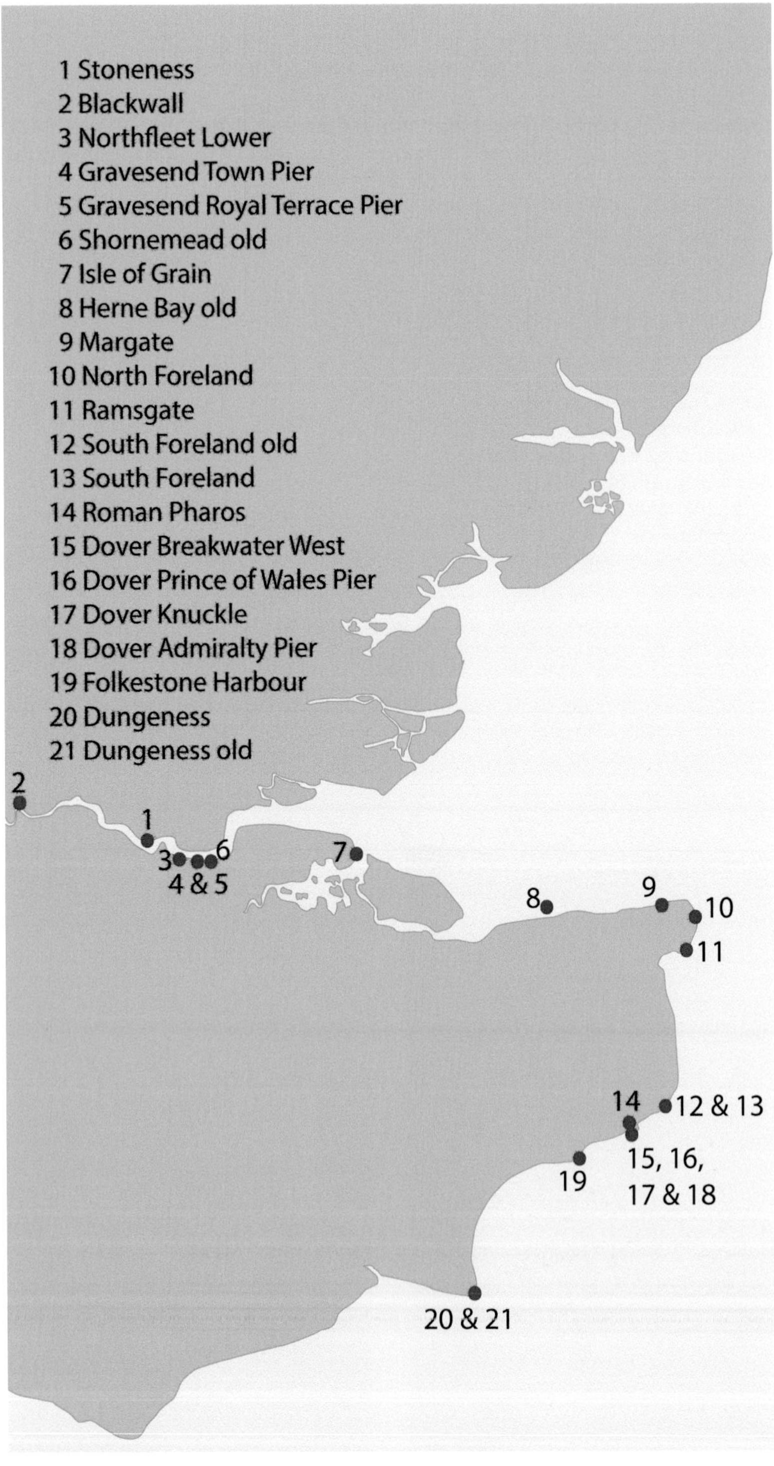
1 Stoneness
2 Blackwall
3 Northfleet Lower
4 Gravesend Town Pier
5 Gravesend Royal Terrace Pier
6 Shornemead old
7 Isle of Grain
8 Herne Bay old
9 Margate
10 North Foreland
11 Ramsgate
12 South Foreland old
13 South Foreland
14 Roman Pharos
15 Dover Breakwater West
16 Dover Prince of Wales Pier
17 Dover Knuckle
18 Dover Admiralty Pier
19 Folkestone Harbour
20 Dungeness
21 Dungeness old
2
1
3
4 & 5
6
7
8
9
10
11
14
12 & 13
15, 16,
17 & 18
19
20 & 21

Region 20

Thames Estuary And Kent

20.1 STONENESS (t) ☆

North bank of Thames at West Thurrock

Red metal framework tower (13 metres)
Established: 1885
Admiralty number: A2126
Light characteristic: Fl G 2.5s
Grid reference: TQ 586 760
Access inside tower: No

Access to site: The lighthouse can be seen from the coastal path at West Thurrock. To access the coastal path take the footpath off of the A126, London Road, next to St Clement's Church in West Thurrock. Once at the river head west. The walk is around 1.5km (1 mile) in total from the main road to the best vantage point.

20.2 BLACKWALL (BOW CREEK, NORTH LEE) (m)

In Trinity Buoy Wharf, south west of Leamouth ☆

Six-sided brick tower, black gallery, white lantern
Established: 1846
Designed by: Sir James Douglass

○ Visited on ....

Light never routinely operational
Grid reference: TQ 395 808
Access inside tower: Yes. The lighthouse is open to the public for self-guided tours.

Access to site: The lighthouse is accessible on foot from the end of Orchard Place, south of Canning Town, through the entrance to Trinity Buoy Wharf.

Note: The former lens from the lighthouse is now on display at Hurst Castle in Hampshire.

20.3 NORTHFLEET LOWER (m) ☆

○ Visited on ....

South bank of Thames at Northfleet

Red metal framework tower, red lantern (16 metres)
Established: 1883
Light discontinued
Grid reference: TQ 625 747
Access inside tower: No

Access to site: The lighthouse is on private land but can be seen from the end of The Shore road in Northfleet.

20.4 GRAVESEND TOWN PIER (m)

○ Visited on ....

On Town Pier ☆

White tower with lantern rising from single storey building
Light discontinued
Grid reference: TQ 648 745
Access inside tower: No

Access to site: There is public access to the pier from which the light is visible. A good starting point is the Market Square or Horn Yard car parks off of Bank Street. From here a walk along the river takes in the Town Pier light as well as Light Vessel 21 and the Royal Terrace Pier light (see 20.5).

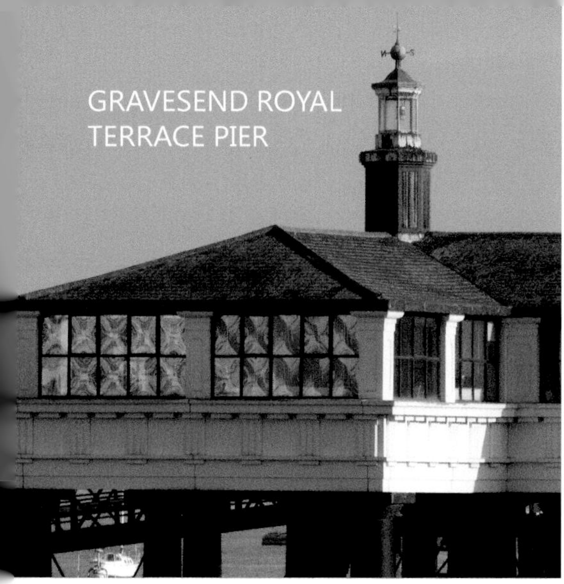

20.5 GRAVESEND ROYAL TERRACE PIER (m)

On Royal Terrace Pier ☆

Stone tower with cupola lantern rising from single storey building
Admiralty number: A2098
Light characteristic: F R
Grid reference: TQ 651 745
Access inside tower: No

○ Visited on ....

Access to site: The lighthouse is visible just beyond the entrance to Royal Terrace Pier from Royal Pier Road and should not be confused with the structure that looks similar to a lighthouse at the pier entrance. There is no public access to the pier. A good starting point is the Market Square or Horn Yard car parks off of Bank Street. From here a walk along the river takes in the Royal Terrace Pier light as well as Light Vessel 21 and the Town Pier light (see 20.4).

20.6 SHORNEMEAD OLD (m) ☆

○ Visited on ....

Relocated a short distance from its original location to Denton Wharf, east of Gravesend

Red cylindrical metal tower on red piles, red lantern (14 metres)
Established: 1913
Light discontinued 2004
Grid reference: TQ 665 743
Access inside tower: No

Access to site: The lighthouse is in a private enclosure, but clearly visible on Mark Lane in the Milton area of Gravesend.

20.7 ISLE OF GRAIN (m)

○ Visited on ....

East coast of Isle of Grain

Skeletal tower on small hut (20 metres)
Admiralty number: A2009.7

Light characteristic: F WRG
Grid reference: TQ 893 753
Access inside tower: No

Access to site: The lighthouse is accessible on foot from the end of the public section of Port Victoria Road where there is a parking area. From here it is a 1km (0.6 mile) walk south towards the coast and then south to reach the lighthouse. If following the road on foot from the parking area, the light can be found by turning left at the hardstanding just beyond the fenced-off area.

20.8 HERNE BAY OLD (i)

On remains of Herne Bay Pier

○ Visited on ....

Concrete building on concrete piles, octagonal lantern (15 metres)
Light discontinued
Grid reference: TR 169 693
Access inside tower: No

Access to site: There is no access to the site, but the lighthouse and remains of the pier can be seen along the seafront at Herne Bay.

20.9 MARGATE (m) ✯

On end of Margate pier

○ Visited on ....

Stone tower (20 metres)
Established: 1955
Admiralty number: A1972
Light characteristic: F R
Grid reference: TR 351 712
Access inside tower: No

Access to site: The pier is accessible from Rendezvous, with parking available at the pier entrance.

20.10 NORTH FORELAND (m)

North of Broadstairs ✯

White octagonal tower (26 metres)
Established: 1691, modernised 1790
Admiralty number: A0966
Light characteristic: Fl(5)WR 20s
Grid reference: TR 398 697
Access inside tower: No

○ Visited on ....

Access to site: The lighthouse is at the side of North Foreland Road. Access inside to compound is for those staying at the keepers' cottages only.

Note: North Foreland was the last of the manned lighthouses operated by Trinity House, with the last keepers leaving in 1998.

20.11 RAMSGATE (m) ✯

○ Visited on ....

West pier head in Ramsgate

Round granite tower (11 metres)
Established: 1842
Admiralty number: A0957.1
Light characteristic: F R
Grid reference: TR 384 643
Access inside tower: No

Access to site: The west pier and lighthouse are accessible on foot from Military Road. There is parking a short distance from the entrance to the pier.

20.12 SOUTH FORELAND OLD (LOWER SOUTH FORELAND) (m) ✯

○ Visited on ....

East of South Foreland lighthouse

White octagonal tower with lantern (15 metres)
Established: 1793
Light discontinued 1904
Grid reference: TR 362 434
Access inside tower: No

Access to site: The lighthouse and surrounding area are now privately owned and there is no public access. The lighthouse is best seen from the top of South Foreland lighthouse (see 20.13) or the coastal path to the south east.

20.13 SOUTH FORELAND (m) ✯

○ Visited on ....

South of St Margaret's-at-Cliffe

White octagonal tower with lantern (21 metres)
Established: 1843
Designed by: James Walker
Light discontinued 1988
Grid reference: TR 359 433
Access inside tower: Yes. The lighthouse is open for guided tours from February to late October, and at weekends during the winter months. It is advisable to check the National Trust website for opening times before travelling if you wish to enter the tower.

Access to site: There is no vehicular access to the lighthouse. Access is on foot on a 2-mile cliff top walk eastbound from the National Trust White Cliffs of Dover car park. The car park is just off of Upper Road and the White Cliffs are signposted from the A258, Castle Hill Road in Dover. The four lighthouses in Dover harbour (see 20.15, 20.16, 20.17 and 20.18) are visible during this walk.

20.14 ROMAN PHAROS (DUBRIS PHAROS, ROMAN LIGHTHOUSE) (m)

○ Visited on ....

Within Dover Castle

Six-sided stone tower (19 metres)
Established: 46-50AD
Light discontinued
Grid reference: TR 325 419
Access inside tower: Yes. It is possible to enter the ground floor only of the lighthouse.

Access to site: The tower is next to St Mary's Church within the grounds of Dover Castle.

20.15 DOVER BREAKWATER WEST (i) ✯

○ Visited on ....

West end of breakwater, Dover

White tower (21 metres)
Established: 1909

Admiralty number: A0924
Light characteristic: OC R 30s
Grid reference: TR 331 401
Access inside tower: No

Access to site: There is no public access to the breakwater. The lighthouse can be seen from ferries out of/arriving at Dover, from the seafront, or at a distance from the cliff top walk towards South Foreland lighthouse (see 20.13).

20.16 DOVER PRINCE OF WALES PIER (m) ✯

End of inner south pier, Dover

◯ Visited on ....

White tower (14 metres)
Established: 1902
Admiralty number: A0902
Light characteristic: Fl Y 1.5s
Grid reference: TR 326 402
Access inside tower: No

Access to site: There is no longer public access to the pier. The lighthouse can be seen from ferries out of/arriving at Dover, from the seafront, or at a distance from the cliff top walk towards South Foreland lighthouse (see 20.13).

20.17 DOVER KNUCKLE (i) ✯

Part way along breakwater, Dover

◯ Visited on ....

White tower (16 metres)
Established: 1909
Admiralty number: A0926
Light characteristic: Fl(4)WR 10s
Grid reference: TR 339 406
Access inside tower: No

Access to site: There is no public access to the breakwater. The lighthouse can be seen from ferries out of Dover, from the seafront, or at a distance from the cliff top walk towards South Foreland lighthouse (see 20.13).

20.18 DOVER ADMIRALTY PIER (m) ✯

On end of south pier, Dover

○ Visited on ....

White tower (22 metres)
Established: 1908
Admiralty number: A0900
Light characteristic: Fl W 7.5s
Grid reference: TR 330 399
Access inside tower: No

Access to site: There is no public access to the pier. The lighthouse can be seen from ferries out of/arriving at Dover, from the seafront, or at a distance from the cliff top walk towards South Foreland lighthouse (see 20.13).

20.19 FOLKESTONE HARBOUR (m) ✯

On end of Folkestone pier

○ Visited on ....

Tower (13 metres)
Established: 1860
Admiralty number: A0892
Light characteristic: Fl(2)W 10s
Grid reference: TR 238 356
Access inside tower: Yes. The lighthouse is now open as a Champagne bar on Fridays and weekends.

Access to site: The pier is accessible on foot from the car park at the pier entrance, off of Harbour Approach.

20.20 DUNGENESS (m) ✯

○ Visited on ....

East of Dungeness Power Station

White round tower, black bands, white lantern (43 metres)
Established: 1961

Admiralty number: A0876
Light characteristic: Fl W 10s
Grid reference: TR 093 169
Access inside tower: No

Access to site: Dungeness can be accessed by heading south from Lydd along Dungeness Road, or along the coastal road south of Greatstone. The modern lighthouse is accessible on foot from the old lighthouse (see 20.21). Parking is opposite the old lighthouse.

20.21 DUNGENESS OLD (HIGH LIGHT TOWER) (m) ☆

○ Visited on ....

East of Dungeness Power Station

Black tower, white lantern (43 metres)
Established: 1904
Light discontinued 1961
Grid reference: TR 088 169
Access inside tower: Yes. The lighthouse is open routinely however it is best to check the Dungeness Lighthouse website for opening times before travelling. The lighthouse is closed during the winter months.

Access to site: Dungeness can be accessed by heading south from Lydd along Dungeness Road, or along the coastal road south of Greatstone. Parking is opposite the old lighthouse. The modern lighthouse (see 20.20) is close by and accessible on foot.

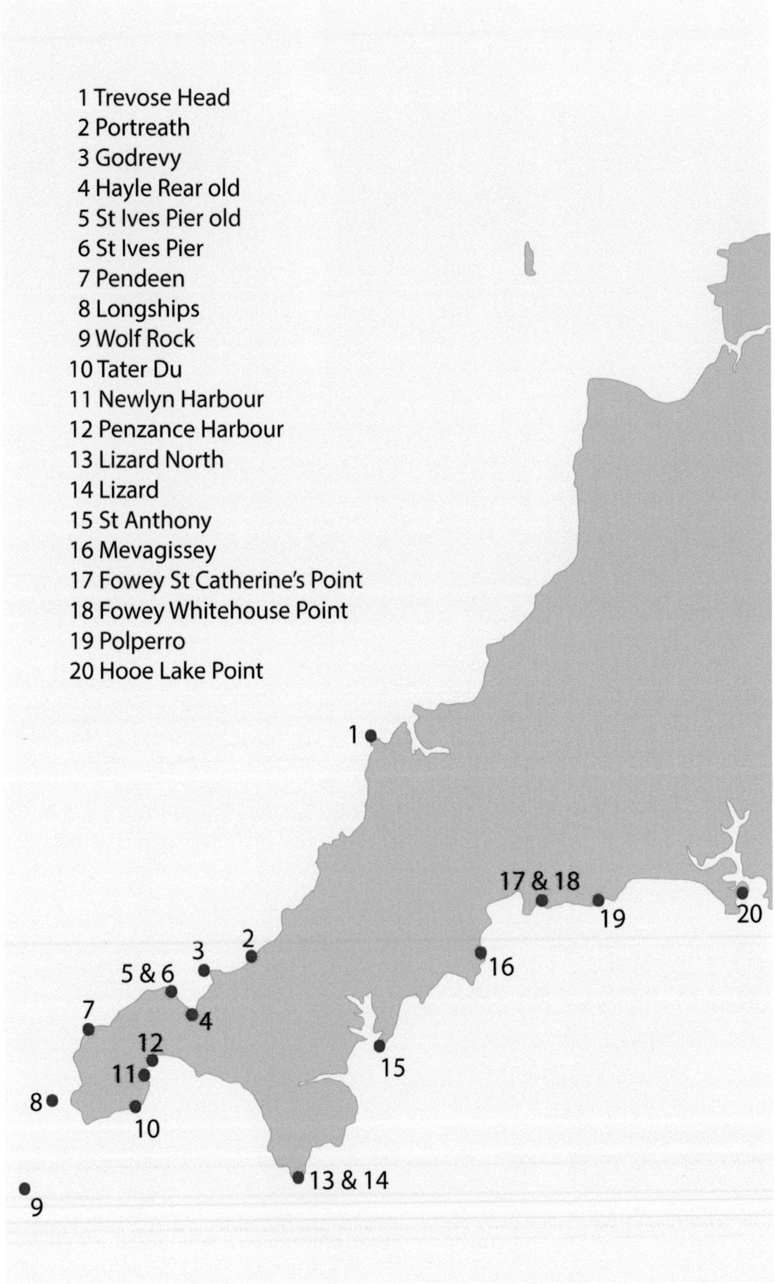
1 Trevose Head
2 Portreath
3 Godrevy
4 Hayle Rear old
5 St Ives Pier old
6 St Ives Pier
7 Pendeen
8 Longships
9 Wolf Rock
10 Tater Du
11 Newlyn Harbour
12 Penzance Harbour
13 Lizard North
14 Lizard
15 St Anthony
16 Mevagissey
17 Fowey St Catherine's Point
18 Fowey Whitehouse Point
19 Polperro
20 Hooe Lake Point
1
17 & 18
19
20
2
3
16
5 & 6
7
4
12
15
11
8
10
13 & 14
9

Region 21

Cornwall

21.1 TREVOSE HEAD (m) ☆

North west of Constantine Bay

White round tower (27 metres)
Established: 1847
Designed by: Sir James Douglass
Admiralty number: A5638
Light characteristic: Fl W 7.5s
Grid reference: SW 851 766
Access inside tower: No

○ Visited on

....

Access to site: From the village of Trevose follow Trevose Head, which becomes a toll road. Parking is available just before you reach the lighthouse at the end of the road.

21.2 PORTREATH (MONKEY HUT) (m)

At end of Landmark Pier, Portreath

○ Visited on ....

White cylindrical stone tower (6 metres)
Established: 1812, re-built 2014/15
Light discontinued
Grid reference: SW 654 456
Access inside tower: No

Access to site: There is no public access to the pier. The best vantage points are from the daymark which is a short walk from Lighthouse Hill, or from Portreath beach.

21.3 GODREVY (i) ✭

Off of Godrevy Point

○ Visited on ....

White octagonal stone tower (26 metres)
Established: 1859
Designed by: James Walker
Admiralty number: A5654
Light characteristic: Fl WR 10s
Grid reference: SW 577 436
Access inside tower: No

Access to site: The island is privately owned and landing is not permitted. The lighthouse can be seen from the South West Coast Path to the north of Gwithian. From Gwithian follow the B3301 north and then follow signs to Godrevy. Parking is available close to the viewpoint. Boat trips around the island run from St Ives during the summer months.

21.4 HAYLE REAR OLD (m)

North of Lelant

○ Visited on ....

Pile structure, red and white lantern (5 metres)
Established: 1840
Admiralty number: A5658.1
Light characteristic: F W

Grid reference: SW 548 379
Access inside tower: No

Access to site: The lighthouse is accessible on foot along the South West Coast Path from the corner of Church Road and Church Lane in Lelant, where parking is available. From here it is a short walk to the lighthouse, which is to the north west just before the path passes under the railway line.

21.5 ST IVES PIER OLD (SMEATON'S PIER OLD) (m) ☆

○ Visited on ....

Midway along Smeaton's pier, St Ives

Square stone base, octagonal stone gallery
Established: 1830
Light discontinued 1890
Grid reference: SW 521 407
Access inside tower: No

Access to site: The pier is accessible from Quay Street. Parking is often difficult in St Ives. There is limited parking on the pier and a public car park at the end of Island Road, but the park and ride may be a more suitable option at busier times of year. The operational light (see 21.6) is a little further along the pier.

21.6 ST IVES PIER (SMEATON'S PIER) (m) ☆

At end of Smeaton's pier, St Ives

○ Visited on ....

White round metal tower (10 metres)
Established: 1890
Admiralty number: A5665
Light characteristic: 2 F G(vert)
Grid reference: SW 521 406
Access inside tower: No

Access to site: The pier is accessible from Quay Street. Parking is often difficult in St Ives. There is limited parking on the pier and a public car park at the end of Island Road, but the park and ride may be a more suitable option at busier times of year. You will pass the old light (see 21.5) on your way along the pier.

HAYLE REAR OLD

ST IVES PIER

21.7 PENDEEN (PENDEEN WATCH) (m) ✫

North of Pendeen

○ Visited on ....

White round tower (17 metres)
Established: 1900
Designed by: Thomas Matthews
Admiralty number: A5670
Light characteristic: Fl(4)W 15s
Grid reference: SW 379 359
Access inside tower: No

Access to site: The lighthouse is accessible by following Calartha Road north off of the main road through Pendeen village. There is a parking area just beyond the lighthouse.

21.8 LONGSHIPS (r) ✫

2km off of Land's End

○ Visited on ....

Grey round granite tower (35 metres)
Established: 1875
Designed by: Sir James Douglass
Admiralty number: A0028
Light characteristic: Fl(2)WR 10s
Grid reference: SW 320 253
Access inside tower: No

Access to site: The lighthouse is visible from Land's End. A closer view would require a chartered boat.

21.9 WOLF ROCK (r) ✫

13.7km off of Gwennap Head

○ Visited on ....

Grey round granite tower, black lantern (41 metres)
Established: 1869
Designed by: James Walker
Admiralty number: A0030
Light characteristic: Fl W 15s
Grid reference: SW 268 120
Access inside tower: No

Access to site: The lighthouse is best seen by chartered boat, but it is visible from a distance on a clear day from Gwennap Head. The tower can also be seen from the Scillonian ferry between Penzance and the Isles of Scilly.

Note: An iron conical beacon, which pre-dates the lighthouse, can be seen next to the lighthouse when in closer proximity.

21.10 TATER DU (m)

○ Visited on ....

South west of Lamorna ☆

White round tower (15 metres)
Established: 1965
Designed by: Michael H Crisp
Admiralty number: A0032
Light characteristic: Fl(3)W 15s
Grid reference: SW 440 230
Access inside tower: No

Access to site: Access to the lighthouse compound is prohibited, but the lighthouse is visible from the South West Coast Path. Parking is available at Lamorna Cove and from here it is 1.8km (1 mile) walk south west along a path to the lighthouse. The path is uneven in places making the walk a little strenuous.

21.11 NEWLYN HARBOUR (m) ☆

○ Visited on ....

At end of south pier, Newlyn

White round metal tower, red base and cupola (10 metres)
Established: 1915
Admiralty number: A0038
Light characteristic: Fl W 5s
Grid reference: SW 468 286
Access inside tower: No

Access to site: The pier is closed to the public however the light can be seen from the north pier, which is accessible on foot from the Strand. Parking is available at Duke Street car park on the corner of the Strand and Jack Lane, a short walk away.

21.12 PENZANCE HARBOUR (m) ☆

○ Visited on ....

At end of south pier, Penzance

White round tower, black base (9 metres)
Established: 1855
Admiralty number: A0046
Light characteristic: Fl WR 5s
Grid reference: SW 479 301
Access inside tower: No

Access to site: The pier is accessible on foot from Battery Road. Parking is available off of Green Street or on Western Promenade Road.

21.13 LIZARD NORTH (m)

◯ Visited on ....

At Lizard Point, the most southerly point of mainland Britain

White octagonal tower at north end of building, lantern removed (19 metres)
Established: 1751
Light discontinued 1903
Grid reference: SW 704 116
Access inside tower: There is access to some parts of the tower for those staying in one of the holiday cottages.

Access to site: The lighthouse is at the end of Lighthouse Road, to the south of Lizard village. There is a public car park close to the lighthouse. The operational lighthouse (see 21.14) is within the same compound.

21.14 LIZARD (m) ✯

◯ Visited on ....

At Lizard Point, most southerly point of mainland

White octagonal tower at south end of building, white lantern (19 metres)
Established: 1751
Designed by: Thomas Fonnereau
Admiralty number: A0082
Light characteristic: Iso WR 15s
Grid reference: SW 704 115
Access inside tower: Yes. The Visitor Centre, including tours of the tower, is open Sunday to Thursday from late March to late October, although it is best to check on the Trinity House website before travelling if you wish to enter the lighthouse.

Access to site: The lighthouse is at the end of Lighthouse Road, to the south of Lizard village. There is a public car park close to the lighthouse. The north tower (see 21.13) is within the same compound.

21.15 ST ANTHONY (m) ✯

◯ Visited on ....

On the southern tip of the Roseland Peninsula

White octagonal tower (19 metres)
Established: 1835
Designed by: James Walker
Admiralty number: A0062

LIZARD NORTH

LIZARD

ST ANTHONY

Light characteristic: Iso WR 15s
Grid reference: SW 846 311
Access inside tower: No

Access to site: The lighthouse is accessible by a short walk from the car park at the end of Military Road, to the south east of Bohortha village.

21.16 MEVAGISSEY (m) ✯

At the end of the south pier in Mevagissey

○ Visited on ....

White metal tower, black base (8 metres)
Established: 1896
Admiralty number: A0078
Light characteristic: Fl(2)W 10s
Grid reference: SX 018 447
Access inside tower: No

Access to site: The pier is accessible on foot and there is public parking at the entrance to the pier.

21.17 FOWEY ST CATHERINE'S POINT (m) ✯

At west entrance to River Fowey

○ Visited on ....

Octagonal platform, red lantern (6 metres)
Established: 1904
Admiralty number: A0082
Light characteristic: LFl WR 5s
Grid reference: SX 117 508
Access inside tower: No

Access to site: The lighthouse is behind a locked gate off of the South West Coast Path to the south west of Fowey. A distant view can be gained close to St Saviours Hill in Polruan.

Note: The Trinity House Gribben Head daymark can be clearly seen to the south west from both the South West Coast path in the area and from Polruan.

21.18 FOWEY WHITEHOUSE POINT (m)

Near departure point for ferry to Polruan

○ Visited on ....

Red metal column (4 metres)
Established: 1904

MEVAGISSEY

FOWEY

FOWEY WHITEHOUSE POINT

Admiralty number: A0084
Light characteristic: Iso WRG 3s
Grid reference: SX 124 515
Access inside tower: No

Access to site: The light is accessible on foot from the Esplanade at the end of the lane signposted Polruan ferry.

21.19 POLPERRO (m)

○ Visited on ....

East of Polperro harbour

White concrete pillar (3 metres)
Established: 1911
Admiralty number: A0090
Light characteristic: Iso WR 6s
Grid reference: SX 213 508
Access inside tower: No

Access to site: The lighthouse is accessible on foot from the large car park on the A387 just before entering Polperro village. Parking in the village is not permitted. From the car park, walk to the harbour and then follow The Warren along the east side of the harbour. In front of the final house on The Warren a small track to the south will take you to the lighthouse. It is a 1.2km (0.75 mile) walk from the car park to the lighthouse. Alternatively, the light can be approached from the east by parking at Talland Bay. From this side it is a 1.6km (1 mile) walk to the lighthouse along the South West Coast Path before you will see the small track leading down in front of the first house you see on the approach to Polperro.

21.20 HOOE LAKE POINT (m)

○ Visited on ....

East of Maker

White concrete tower, red stripe (5 metres)
Admiralty number: A0112
Light characteristic: Fl(2)WRG 10s
Grid reference: SX 448 514
Access inside tower: No

Access to site: The lighthouse is within the Hooe Lake Campsite towards the end of the public road to the south east of Maker. The road is signposted 'Fort Picklecombe' from Earl's Drive in Maker.

POLPERRO

HOOE LAKE POINT

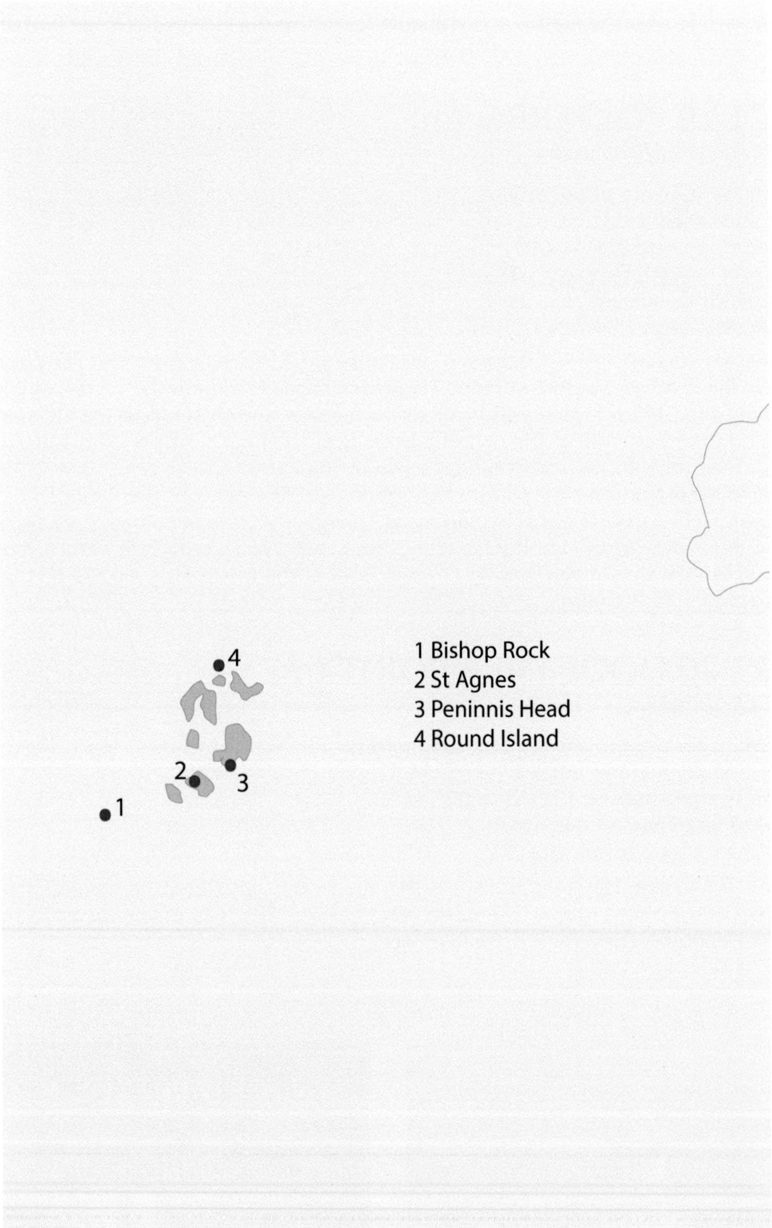
1 Bishop Rock
2 St Agnes
3 Peninnis Head
4 Round Island

Region 22

Isles Of Scilly

22.1 BISHOP ROCK (r) ✯

5.4km south west of Annet island

Grey round granite tower (49 metres)
Established: 1858, strengthened 1887
Designed by: James Walker / Sir James Douglass
Admiralty number: A0002
Light characteristic: Fl(2)W 15s
Grid reference: SV 808 065
Access inside tower: No

○ Visited on

....

Access to site: Sightseeing boat trips from both St Mary's and St Agnes sail to the lighthouse during the summer months, weather permitting. Alternatively, the lighthouse can be seen on a clear day from St Agnes, which is accessible by scheduled ferry from St Mary's island.

22.2 ST AGNES (i) ✯

Centre of St Agnes island

◯ Visited on ....

White round tower (15 metres)
Established: 1680
Designed by: Capt. Hugh Till and Capt. Symon Bayly
Light discontinued 1911
Grid reference: SV 881 082
Access inside tower: No

Access to site: St Agnes is accessible by scheduled ferry from St Mary's island. The lighthouse is visible from the main path through the village on St Agnes.

22.3 PENINNIS HEAD (i) ✯

Southern Point of St Mary's island

◯ Visited on ....

White round metal tower on black framework tower, black cupola (14 metres)
Established: 1911
Admiralty number: A0006
Light characteristic: Fl W 20s
Grid reference: SV 911 094
Access inside tower: No

Access to site: The lighthouse is accessible on foot from Church Road (A3112) southbound past the hospital and health centre. The walk to the lighthouse from the hospital is around 1km (0.6 miles) on a road and track. Alternatively, the lighthouse can be reached on the coastal path from either Porthcressa Beach or Old Town Bay.

22.4 ROUND ISLAND (i) ✯

On Round island

◯ Visited on ....

White round tower (19 metres)
Established: 1887
Designed by: William Tregarthen Douglass
Admiralty number: A0018
Light characteristic: Fl W 10s
Grid reference: SV 902 177
Access inside tower: No

Access to site: The island and lighthouse are visible from the north coasts of Tresco and St Martin's, both can be reached by scheduled ferry from St Mary's island. A closer view would require a chartered boat.

PENINNIS HEAD

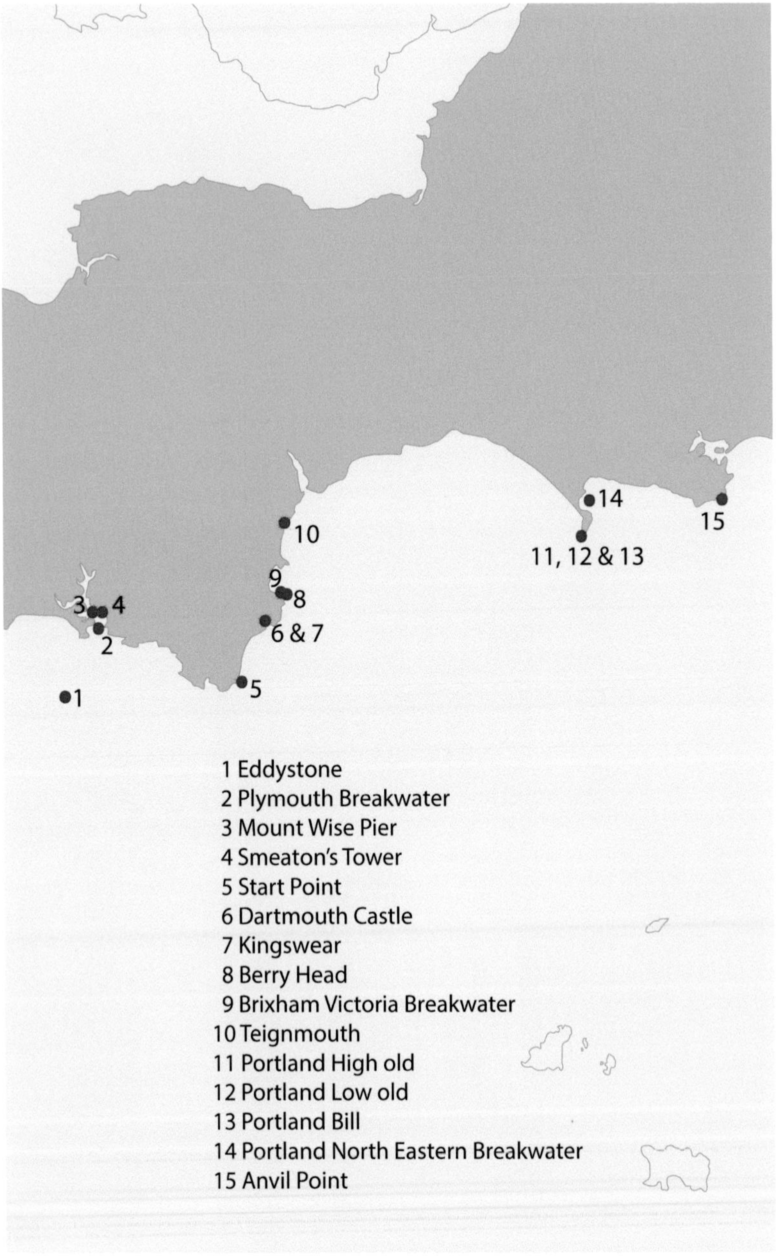

1
2
3
4
5
6 & 7
8
9
10
11, 12 & 13
14
15
1 Eddystone
2 Plymouth Breakwater
3 Mount Wise Pier
4 Smeaton's Tower
5 Start Point
6 Dartmouth Castle
7 Kingswear
8 Berry Head
9 Brixham Victoria Breakwater
10 Teignmouth
11 Portland High old
12 Portland Low old
13 Portland Bill
14 Portland North Eastern Breakwater
15 Anvil Point

Region 23

South Devon And Dorset

23.1 EDDYSTONE (r) ✯

14.6km south of Rame Head

Grey granite tower, red lantern (49 metres)
Established: 1882
Designed by: Sir James Douglass
Admiralty number: A0098
Light characteristic: Fl(2)W 10s
Grid reference: SX 383 336
Access inside tower: No

○ Visited on ....

Access to site: Best seen by chartered boat. The tower can be seen at a great distance on a very clear day from elevated positions around Plymouth. The base of the previous lighthouse (see 23.4) can still be seen at the site.

23.2 PLYMOUTH BREAKWATER (i) ✯ ◯ Visited on ....

On west end of Plymouth Breakwater

White round granite tower (23 metres)
Established: 1844
Admiralty number: A0114
Light characteristic: Fl WR 10s
Grid reference: SX 464 505
Access inside tower: No

Access to site: The lighthouse can be seen from Plymouth, although the closest vantage point would be from the west at the end of the public road towards Fort Picklecombe to the east of Maker, just beyond Hooe Lake Point (see 21.20 for directions). Ferries to and from Plymouth pass close to the lighthouse. A closer view would require a chartered boat.

23.3 MOUNT WISE PIER (m) ◯ Visited on ....

South east of Devonport

White triangle, red stripe on white hut
Admiralty number: A0172
Light characteristic: Dir WRG
Grid reference: SX 456 540
Access inside tower: No

Access to site: The light is at the entrance of Mount Wise pier, which can be accessed on foot from Mount Wise Park. Parking is available on the coastal section of George Street near the Scott Memorial. From here walk down the steps behind the statue or use the sloped path until you reach the bottom of the steps. Continue down the small set of steps opposite and through the arch. The light is straight ahead.

23.4 SMEATON'S TOWER (m) ✯ ◯ Visited on ....

Relocated to Plymouth Hoe from location of current Eddystone lighthouse

Red and white granite tower (22 metres)

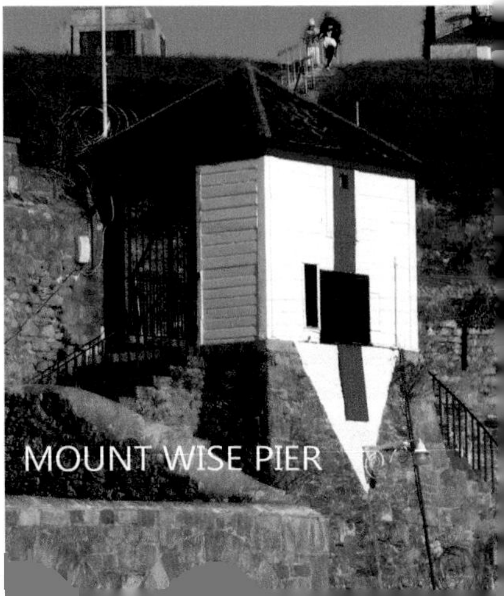

Established: 1759
Designed by: John Smeaton
Light discontinued 1882
Grid reference: SX 478 538
Access inside tower: Yes. The tower is open daily for self-guided tours from 10am to 5pm.

Access to site: Parking bays are available on Hoe Road, a short walk from the lighthouse. The Elphinstone car park on Madeira Road is also nearby.

23.5 START POINT (m) ✯

○ Visited on ....

South west of Dartmouth

White round granite tower (28 metres)
Established: 1836
Designed by: James Walker
Admiralty number: A0228
Light characteristic: Fl(3)W 10s
Grid reference: SX 829 371
Access inside tower: Yes. The lighthouse has a visitor centre which runs guided tours of the tower during the summer months.

Access to site: The turning towards Start Point is signposted from the main road in Bickerton. Parking is available at the end of the road. From the car park it is a 1km (0.6 mile) walk on a tarmac road to the lighthouse.

23.6 DARTMOUTH CASTLE (m)

○ Visited on ....

In the grounds of Dartmouth Castle

White square tower
Established: 1856
Light discontinued
Grid reference: SX 886 503
Access inside tower: No

Access to site: Access to the castle is along Castle Road from the B3205, south of Dartmouth.

23.7 KINGSWEAR (m)

○ Visited on

....

East of Dartmouth

White round GRP tower
Established: 1981
Admiralty number: A0236
Light characteristic: Iso WRG 3s
Grid reference: SX 885 508
Access inside tower: No

Access to site: The beach on which the lighthouse is located is now closed to the public. Depending on the condition of the land, it may be possible to see the lighthouse from the end of Beacon Road (no vehicular access), off of Church Hill. Alternatively, the light can be seen from a distance to the south across the river, from Castle Road on the approach to Dartmouth Castle and from St Petrox Church.

23.8 BERRY HEAD (m) ✯

○ Visited on

....

On headland east of Brixham

White tower (5 metres)
Established: 1906
Admiralty number: A0244
Light characteristic: Fl(2)W 15s
Grid reference: SX 946 566
Access inside tower: No

Access to site: Parking for Berry Head is available at the junction of Berry Head Road and Gillard Road, Brixham. From here it is a 0.8km (0.5 mile) walk to the lighthouse.

23.9 BRIXHAM VICTORIA BREAKWATER (m) ☆

○ Visited on ....

On end of breakwater to north east of Brixham

White tower (6 metres)
Established: 1916
Admiralty number: A0246
Light characteristic: Oc R 15s
Grid reference: SX 925 573
Access inside tower: No

Access to site: The breakwater is accessible on foot. Parking is available at the entrance to the breakwater off of Berry Head Road.

23.10 TEIGNMOUTH (m) ☆

○ Visited on ....

Near north entrance to River Teign

Grey round stone tower (6 metres)
Established: 1845
Admiralty number: A0262
Light characteristic: F R
Grid reference: SX 941 725
Access inside tower: No

Access to site: The lighthouse is next to the car park at the end of Powderham Terrace.

23.11 PORTLAND HIGH OLD (m) ☆

○ Visited on ....

North of Portland Bill

White round tower, lantern and gallery (12 metres)
Established: 1869
Light discontinued 1906
Grid reference: SY 677 692
Access inside tower: Access to the tower is available for those staying in the adjoining cottage.

Access to site: The lighthouse is at the end of Higher Light Lane, Portland Bill. Both the old low light (see 23.12) and the operational light (see 23.13) are nearby.

23.12 PORTLAND LOW OLD (m) ☆

○ Visited on ....

North east of Portland Bill

White round tower (25 metres)
Established: 1869
Light discontinued 1906
Grid reference: SY 681 689
Access inside tower: The lighthouse now contains the Portland Bird Observatory, including accommodation.

Access to site: The lighthouse is at the side of Portland Bill Road, on the approach to the operational Portland Bill lighthouse (see 23.13). The old high lighthouse (see 23.11) is also nearby.

23.13 PORTLAND BILL (m) ☆

○ Visited on ....

Southern point of Portland Bill

White round tower, red band (41 metres)
Established: 1906
Admiralty number: A0294
Light characteristic: Fl(4)W 20s
Grid reference: SY 678 684

Access to tower: The Visitor Centre and lighthouse are open to the public regularly throughout the year. Opening dates and times can be found on the Trinity House website.

Access to site: The lighthouse is at the end of Portland Bill Road with public parking nearby. The old high (see 23.11) and old low (see 23.12) lighthouses are nearby.

23.14 PORTLAND NORTH EASTERN BREAKWATER (i) ✯

○ Visited on ....

On south east end of north eastern breakwater

White metal tower (22 metres)
Established: 1905
Admiralty number: A0314
Light characteristic: Fl W 2.5s
Grid reference: SY 705 763
Access inside tower: No

Access to site: The lighthouse is visible from a distance from Portland Beach Road. A closer view would require a chartered boat.

23.15 ANVIL POINT (m) ✯

○ Visited on ....

Within Durlston Country Park near Swanage

White round tower (12 metres)
Established: 1881
Admiralty number: A0496
Light characteristic: Fl W 10s
Grid reference: SZ 029 769
Access inside tower: No

Access to site: The lighthouse is within Durlston Country Park. Car parking is available at the end of Lighthouse Road, Swanage. From here it is a 0.8km (0.5 mile) walk to the lighthouse.

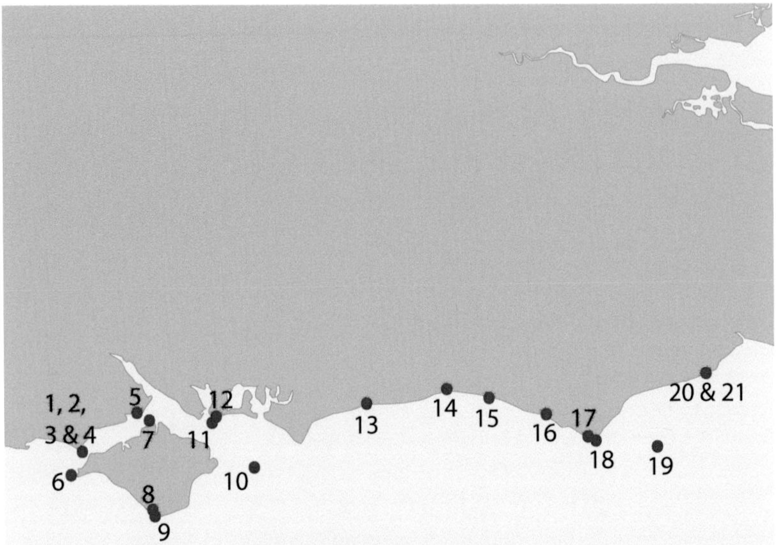

1 Hurst Point
2 Hurst Point Low old
3 Hurst Point old
4 Nab Tower old
5 Beaulieu Millennium Beacon
6 Needles
7 Egypt Point
8 St Catherine's Oratory
9 St Catherine's
10 Nab Tower
11 Spitbank Fort
12 Southsea Castle
13 Littlehampton
14 Shoreham Rear
15 Brighton West Pier
16 Newhaven Breakwater
17 Belle Tout
18 Beachy Head
19 Royal Sovereign
20 Hastings West Hill
21 Hastings Lower Light

Region 24

Hampshire, Isle Of Wight And Sussex

24.1 HURST POINT (m) ✯

On shingle spit to south east of Milford on Sea

White round tower (26 metres)
Established: 1867
Admiralty number: A0538.1
Light characteristic: Fl(4)WR 15s
Grid reference: SZ 319 899
Access inside tower: No

○ Visited on ....

Access to site: The lighthouse can be reached by boat from Keyhaven or on foot along the 2.4km (1.5 mile) shingle spit from Milford. Public parking is available on Keyhaven Road for the ferry or Milford seafront for walking across. The old low light (see 24.2), old light (see 24.3) and old Nab Tower light (see 24.4) are within the grounds of Hurst Castle nearby.

24.2 HURST POINT LOW OLD (m) ☆ ○ Visited on

On shingle spit to south east of Milford on Sea ....

Round granite tower
Established: 1866
Light discontinued 1911
Grid reference: SZ 317 898
Access inside tower: No

Access to site: The lighthouse is within the grounds of Hurst Castle. It can be reached by boat from Keyhaven or on foot along the 2.4km (1.5 mile) shingle spit from Milford. Public parking is available on Keyhaven Road for the ferry or Milford seafront for walking across. The operational light (see 24.1) is nearby. The old light (see 24.3) and old Nab Tower light (see 24.4) are also within the castle grounds.

24.3 HURST POINT OLD (m) ☆ ○ Visited on

On shingle spit to south east of Milford on Sea ....

Square grey metal tower
Established: 1911
Light discontinued 1997
Grid reference: SZ 317 898
Access inside tower: Occasional tours of the tower operate.

Access to site: The lighthouse is within the grounds of Hurst Castle. It can be reached by boat from Keyhaven or on foot along the 2.4km (1.5 mile) shingle spit from Milford. Public parking is available on Keyhaven Road for the ferry or Milford seafront for walking across. The operational light (see 24.1) is nearby. The old low light (see 24.2) and old Nab Tower light (see 24.4) are also within the castle grounds.

24.4 NAB TOWER OLD (m) ☆

○ Visited on ....

Relocated to Hurst Castle on shingle spit to south east of Milford Haven from Nab Tower

Red lantern
Established: 1920
Light discontinued 2013
Grid reference: SZ 316 897
Access inside tower: Public access inside the light should be available from 2019.

Access to site: The lighthouse is within the grounds of Hurst Castle. The old light from the tower is also on display here. It can be reached on foot along 2.4km (1.5 mile) shingle spit from Milford. It can also be reached by boat from Keyhaven. Public parking is available on Milford seafront for walking across or on Keyhaven Road for the ferry. The operational light (see 24.1) is nearby. The old low light (see 24.2) and old light (see 24.3) are also within the castle grounds.

24.5 BEAULIEU MILLENNIUM BEACON (m) ☆

South of Blackfield

○ Visited on ....

White round tower, white lantern (8 metres)
Established: 2000
Admiralty number: A0553.15
Light characteristic: Dir Oc WRG 4s
Grid reference: SZ 449 986
Access inside tower: No

Access to site: The lighthouse is at the side of Lepe Road to the west of Lepe Country Park.

NAB TOWER OLD

BEAULIEU MILLENNIUM BEACON

24.6 NEEDLES (r) ✯

○ Visited on ....

Off of most westerly point of Isle of Wight

Round granite tower, red band, red lantern (31 metres)
Established: 1859
Designed by: James Walker
Admiralty number: A0528
Light characteristic: Oc(2)WRG 20s
Grid reference: SZ 289 848
Access inside tower: No

Access to site: Boat trips around the Needles run during the summer months from the bottom of the cliffs at Freshwater Bay. The lighthouse can also be seen beyond the Needles Old Battery, or a more distant view can be achieved from The Needles Car Park at the end of Alum Bay New Road.

24.7 EGYPT POINT (i)

○ Visited on ....

On promenade in Cowes, Isle of Wight

Red post with white gallery on white round building (7.5 metres)
Established: 1897
Light discontinued 1989
Grid reference: SZ 485 966
Access inside tower: No

Access to site: The light is at the side of Egypt Esplanade.

Note: The former lantern from the lighthouse is now on display at Hurst Castle.

24.8 ST CATHERINE'S ORATORY (i)

○ Visited on ....

On St Catherine's Hill

Octagonal tower, pyramidal roof (10.5 metres)
Established: 1328
Light discontinued 1547
Grid reference: SZ 493 773
Access inside tower: Yes, although access is to the ground floor only.

Access to site: Public parking is available at Blackgang Viewpoint car park. From here steps on the opposite side of the road take you on to the grassy hillside. It is a 0.8km (0.5 mile) walk heading north east from the steps to the oratory.

Note: Close to the oratory is the round stone base of an unfinished lighthouse, started in 1785, which may be worth a visit while in the area.

24.9 ST CATHERINE'S (i) ✯

Southern point of Isle of Wight

○ Visited on

....

White octagonal castellated tower (26 metres)
Established: 1837
Designed by: James Walker
Admiralty number: A0774
Light characteristic: Fl W 5s
Grid reference: SZ 498 753
Access inside tower: Yes. The Visitor Centre, including tours of the lighthouse, is open regularly from Easter to October. Visit the Trinity House website for opening dates and times.

Access to site: The lighthouse is at the end of St Catherine's Road. Access is on foot from the end of the public road.

24.10 NAB TOWER (THE NAB) (i)

East of Isle of Wight

○ Visited on

....

White hexagonal tower
Established: 2013
Admiralty number: A0780
Light characteristic: Fl W 10s
Grid reference: SZ 741 859
Access inside tower: No

Access to site: It is not possible to land on the tower. The lighthouse can be seen from a distance from Bembridge beach on the Isle of Wight or closer by chartered boat.

Note: For details of the previous light in this location see 24.4.

24.11 SPITBANK FORT (i) ☆

○ Visited on ....

South west of Southsea

Red round tower with red lantern (7 metres)
Established: 1866
Admiralty number: A0688
Light characteristic: Fl R 5s
Grid reference: SZ 636 971
Access inside tower: No

Access to site: Access to the fort is only permitted for those staying in its luxury accommodation. The lighthouse is visible from ferries to/from Portsmouth, or from a distance from Southsea.

24.12 SOUTHSEA CASTLE (m) ☆

○ Visited on ....

In grounds of Southsea Castle

White stone tower, black band (10 metres)
Established: 1828
Light discontinued 2017
Grid reference: SZ 643 980
Access inside tower: No

Access to site: The light is at Southsea Castle, which is open to the public. Parking for the castle is nearby.

24.13 LITTLEHAMPTON (m)

○ Visited on ....

At entrance to east pier, Littlehampton

White concrete tower
Established: 1946
Admiralty number: A0801.1
Light characteristic: Oc WY 7.5s
Grid reference: TQ 028 013
Access inside tower: No

Access to site: Parking is available on Arun Parade. From here it is a short walk to the south to reach the lighthouse.

24.14 SHOREHAM REAR (m) ☆

○ Visited on ....

Near inner pier, Shoreham

Grey round stone tower (12 metres)
Established: 1846
Admiralty number: A0814.1
Light characteristic: Fl W 10s
Grid reference: TQ 235 049
Access inside tower: No

Access to site: The lighthouse is next to a public car park off of the A259, Brighton Road.

24.15 BRIGHTON WEST PIER (m)

○ Visited on ....

At end of breakwater, Brighton

White round structure, red bands
Admiralty number: A0826
Light characteristic: Q R
Grid reference: TQ 335 027
Access inside tower: No

Access to site: Public parking is available at Black Rock car park off of Madeira Drive, Brighton. A short walk from here takes you to the pier. Access to the pier is for anglers only, but the lighthouse can be seen from the end of the public section of the pier.

24.16 NEWHAVEN BREAKWATER (m) ☆

At end of breakwater, Newhaven

○ Visited on ....

White round concrete tower, red base (14 metres)
Established: 1891
Admiralty number: A0830
Light characteristic: Oc(2)W 10s
Grid reference: TV 452 994
Access inside tower: No

Access to site: The breakwater is not accessible, although the lighthouse can be seen from the car park at the end of Fort Road.

24.17 BELLE TOUT (m) ✯

West of Beachy Head

○ Visited on ....

Round granite tower (14 metres)
Established: 1832
Designed by: William Hallett and James Walker
Light discontinued 1902
Grid reference: TV 564 955
Access inside tower: Access is for those staying at the B&B.

Access to site: Public parking available off of Beachy Head Road. From here it is a short uphill walk to the lighthouse. Beachy Head lighthouse (see 24.18) is visible with care from the cliffs a short distance away.

24.18 BEACHY HEAD (t) ✯

At foot of cliffs south west of Eastbourne

○ Visited on ....

White round tower, red band and lantern (43 metres)
Established: 1902
Designed by: Thomas Matthews
Admiralty number: A0840
Light characteristic: Fl(2)W 20s
Grid reference: TV 583 951
Access inside tower: No

Access to site: The lighthouse can be reached at low spring tides only. You should check tide times carefully before descending at either Birling Gap to the west or Cow Gap to the east. The lighthouse is also visible from the top of the cliffs near the public parking area close to Belle Tout lighthouse (see 24.17).

24.19 ROYAL SOVEREIGN (i) ✯

South east of Eastbourne

○ Visited on ....

White tower, red band on white cabin, on concrete column
Established: 1971
Admiralty number: A0843
Light characteristic: Fl W 20s

Grid reference: TV 719 945
Access inside tower: No

Access to site: Best seen by boat. Local boat operators may run occasional trips out to the Royal Sovereign. The tower is visible from a distance on a clear day from Eastbourne.

24.20 HASTINGS WEST HILL (m)

○ Visited on ....

North of Lifeboat station, Hastings

White 5-sided tower (6 metres)
Established: 1851
Admiralty number: A0858.1
Light characteristic: F R
Grid reference: TQ 824 097
Access inside tower: No

Access to site: The lighthouse is next to the entrance to Smugglers Adventure in Hastings. Parking is available on Priory Road. From here it is a short walk across the path to the lighthouse. The West Hill Lift funicular railway, open daily during the summer and at weekends in winter, takes you from West Hill to the seafront where the Lower Light (see 24.21) can be seen.

24.21 HASTINGS LOWER LIGHT (m)

○ Visited on ....

Close to seafront, Hastings

White 8-sided tower
Established: 1827
Light discontinued
Grid reference: TQ 824 095
Access inside tower: No

Access to site: The light is on the main A259, which runs along the seafront. The tower can be seen opposite the boating lake. Pelham Road car park is a short walk away. The West Hill Lift funicular railway, open daily during the summer and at weekends in winter, takes you from the seafront to West Hill where the operational light (see 24.20) is located.

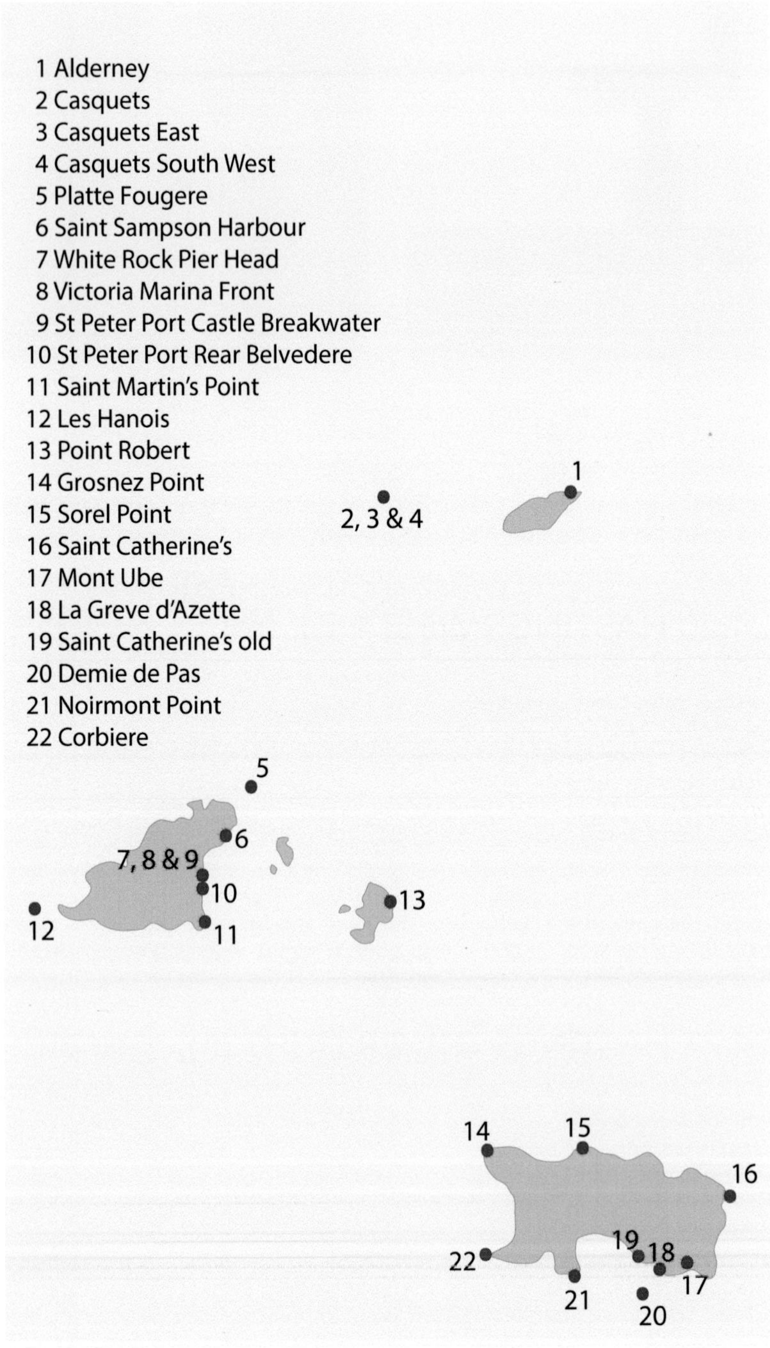
1 Alderney
2 Casquets
3 Casquets East
4 Casquets South West
5 Platte Fougere
6 Saint Sampson Harbour
7 White Rock Pier Head
8 Victoria Marina Front
9 St Peter Port Castle Breakwater
10 St Peter Port Rear Belvedere
11 Saint Martin's Point
12 Les Hanois
13 Point Robert
14 Grosnez Point
15 Sorel Point
16 Saint Catherine's
17 Mont Ube
18 La Greve d'Azette
19 Saint Catherine's old
20 Demie de Pas
21 Noirmont Point
22 Corbiere
1
2, 3 & 4
5
6
7, 8 & 9
10
11
12
13
14
15
16
19
18
22
17
21
20

Region 25

Channel Islands

25.1 ALDERNEY (MANNEZ, QUESNARD) (i) ✭

Northern point of Alderney

White round tower, black band (32 metres)
Established: 1912
Admiralty number: A1536
Light characteristic: Fl(4)W 15s
Lat / long: 49°43.75N 2°09.86W
Access inside tower: Yes. Tours of the lighthouse take place every Sunday from May to September.

◯ Visited on ....

Access to site: It is possible to fly directly to Alderney from Southampton, or you can reach the island by ferry or plane from Guernsey. From Braye harbour on Alderney it is a 3.5km (2 mile) walk to the lighthouse, following the coastal Rue de Beaumont eastbound initially and then heading north east. The lighthouse is next to the road. From the airport it is a 5km (3 mile) walk through St Anne and continuing east.

25.2 CASQUETS (NORTH WEST TOWER, ST PETER) (i) ✯

◯ Visited on ....

West of Alderney, north west tower

White tower, red bands (23 metres)
Established: 1724
Admiralty number: A1532
Light characteristic: Fl(5)W 30s
Lat / long: 49°43.32N 2°22.62W
Access inside tower: No

Access to site: The island and three towers (see also 25.3 and 25.4) can be seen on a clear day from ferries to/from Poole and Portsmouth. A closer view would require a chartered boat.

25.3 CASQUETS EAST (i)

◯ Visited on ....

West of Alderney, east tower

Round stone tower
Established: 1724
Light discontinued 1877
Lat / long: 49°43.32N 2°22.62W
Access inside tower: No

Access to site: The island and three towers (see also 25.2 and 25.4) can be seen on a clear day from ferries to/from Poole and Portsmouth. A closer view would require a chartered boat.

Note: The lighthouse now houses the foghorn equipment.

25.4 CASQUETS SOUTH WEST (i)

◯ Visited on ....

West of Alderney, south west tower

Round white tower with helipad
Established: 1724

Light discontinued 1877
Lat / long: 49°43.32N 2°22.62W
Access inside tower: No

Access to site: The island and three towers (see also 25.2 and 25.3) can be seen on a clear day from ferries to/from Poole and Portsmouth. A closer view would require a chartered boat.

Note: The tower now operates as a helipad.

25.5 PLATTE FOUGÉRE (r)

○ Visited on ....

Off of north east coast of Guernsey

White octagonal tower, black bands (25 metres)
Established: 1910
Admiralty number: A1548
Light characteristic: Fl WR 10s
Lat / long: 49°30.83N 2°29.14W
Access inside tower: No

Access to site: The lighthouse is visible from ferries to/from Poole or Portsmouth. Alternatively, the light can be seen from Fort Doyle, north of Beaucette Marina on Guernsey. A closer view would require a chartered boat.

25.6 SAINT SAMPSON HARBOUR (CROCQ PIERHEAD) (i)

○ Visited on ....

In Saint Sampson harbour, Guernsey

Red column (6 metres)
Admiralty number: A1554
Light characteristic: F R
Lat / long: 49°28.98N 2°31.00W
Access inside tower: No

Access to site: The lighthouse is at the end of the small inner pier. Parking is available on South Quay in Saint Sampson, very close to the lighthouse.

PLATTE FOUGÉRE

SAINT SAMPSON HARBOUR

25.7 WHITE ROCK PIER HEAD (i) ☆

At end of north pier in St Peter Port, Guernsey

○ Visited on ....

Round stone tower (10 metres)
Established: 1908
Admiralty number: A1562
Light characteristic: Oc G 5s
Lat / long: 49°27.38N 2°31.59W
Access inside tower: No

Access to site: There is no public access to the pier. The lighthouse is best seen from ferries arriving at/departing from St Peter Port. It can also be seen from the Castle Breakwater light (see 25.9).

25.8 VICTORIA MARINA FRONT (i) ☆

In Victoria Marina, St Peter Port, Guernsey

○ Visited on ....

White framework tower, red lantern (7 metres)
Admiralty number: A1569
Light characteristic: Oc R 5s
Lat / long: 49°27.32N 2°32.03W
Access inside tower: No

Access to site: The light is accessible at the end of Victoria Pier, where parking is available (with parking disc).

25.9 ST PETER PORT CASTLE BREAKWATER (i)

At end of breakwater in St Peter Port, Guernsey ☆

○ Visited on ....

Dark round granite tower, white on north east side (12 metres)
Established: 1867
Admiralty number: A1560
Light characteristic: Al WR 10s
Lat / long: 49°27.30N 2°31.45W
Access inside tower: No

Access to site: The breakwater is accessible on foot from Castle Pier, where parking is available (with parking disc). From Castle Pier, it is possible to see the Rear Belvedere light (see 25.10) by looking to the south. The White Rock Pier Head (see 25.7) and Victoria Marina Front (see 25.8) lights can be seen from the lighthouse and the pier on the approach.

25.10 ST PETER PORT REAR BELVEDERE (i)

Off of Rue de Belvedere, St Peter Port

◯ Visited on ....

South of St Peter Port, Guernsey
White tower (4 metres)
Admiralty number: A1560.1
Light characteristic: Oc W 10s
Lat / long: 49°26.90N 2°31.98W
Access inside tower: No

Access to site: The lighthouse is between the trees just off of Rue de Belvedere. There is a parking area on the sweeping bend and the entrance to the light is among the trees between here and Belvedere House. The light can also be seen from a distance from Castle Pier in St Peter Port harbour.

◯ Visited on ....

25.11 SAINT MARTIN'S POINT (JERBOURG POINT) (i)

South east point of Guernsey
Flat-roofed white concrete building (5 metres)
Admiralty number: A1574
Light characteristic: Fl(3)WR 10s
Lat / long: 49°25.31N 2°31.70'W
Access inside tower: No

Access to site: The lighthouse is accessible on foot on a path heading south east from the public car park opposite Hotel Jerbourg, off of Route de Jerbourg. It should be noted that during nesting season the area can be populated by aggressive birds.

25.12 LES HANOIS (r) ✯

Off of south west coast of Guernsey

○ Visited on ....

Grey round granite tower, black lantern (33 metres)
Established: 1862
Designed by: Sir James Douglass
Admiralty number: A1580
Light characteristic: Fl(2)W 13s
Lat / long: 49°26.10N 2°42.15W
Access inside tower: No

Access to site: The lighthouse is visible from a distance from the coast of Pleinmont on Guernsey. Boat operators in Guernsey organise regular trips to see the lighthouse in the summer months.

25.13 POINT ROBERT (SARK) (i) ✯

On east coast of Sark

○ Visited on ....

White octagonal tower
Established: 1913
Admiralty number: A1544
Light characteristic: Fl W 15s
Lat / long: 49°26.19N 2°20.75W
Access inside tower: No

Access to site: Sark is accessible by passenger ferry from Guernsey throughout the year, or from Jersey during the summer. The lighthouse is beyond La Valette campsite, which is signposted from the village. It is around 2kms (1.25 miles) from the harbour to the lighthouse and taking the ride on the tractor cart from the harbour to the village is recommended to avoid walking the uphill section. The tower is also visible from the Guernsey ferry on the approach to Maseline harbour on the island.

25.14 GROSNEZ POINT (i)

North west point of Jersey

◯ Visited on ....

White concrete hut
Established: 1948
Admiralty number: A1622
Light characteristic: Fl(2)WR 15s
Lat / Long: 49°15.50N 2°14.80W
Access inside tower: No

Access to site: Public parking is at the end of Le Chemin du Chateau. From here, walk through and beyond the grounds of the castle ruins. A handrail and steps a short distance to the north of the ruins take you down to the lighthouse.

25.15 SOREL POINT (i) ✯

North coast of Jersey

○ Visited on ....

White round concrete tower (3 metres)
Established: 1938
Admiralty number: A1584
Light characteristic: LFl WR 7.5s
Lat / Long: 49°15.60N 2°09.54W
Access inside tower: No

Access to site: Parking is at the end of La Rue de Sorel, north of Sorel. From here it is a short walk to the lighthouse.

25.16 SAINT CATHERINE'S (i)

At end of St Catherine's breakwater, Jersey

○ Visited on ....

White framework tower (9 metres)
Established: 1950
Admiralty number: A1586
Light characteristic: Fl W 1.5s
Lat / Long: 49°13.34N 2°00.65W
Access inside tower: No

Access to site: Public parking (parking disc required) is available at the entrance to the breakwater off of Rte de Ste Catherine.

Note: For details of the former lighthouse in this location see 25.19.

25.17 MONT UBÉ (i) ✯

North of Saint Clement's Bay, Jersey

○ Visited on ....

White metal framework tower (14 metres)
Established: 1896
Admiralty number: A1594.1

Light characteristic: Oc R 5s
Lat / Long: 49°10.30N 2°03.59W
Access inside tower: No

Access to site: The lighthouse is at the side of Rue de la Houguette, north of Le Hocq.

25.18 LA GRÉVE D'AZETTE (i) ✯

○ Visited on ....

South east of St Helier, Jersey

White metal framework tower, white lantern (20 metres)
Established: 1896
Admiralty number: A1594
Light characteristic: Oc W 5s
Lat / Long: 49°10.16N 2°05.09W
Access inside tower: No

Access to site: The lighthouse is on the beach at the side of the A4 near the end of School Road. It is possible to see the Demie de Pas lighthouse (see 25.20) to the south from here.

25.19 SAINT CATHERINE'S OLD (i) ✯

Relocated to St Helier Maritime Museum, Jersey, from location of current Saint Catherine's lighthouse

○ Visited on ....

White six-sided tower
Established: 1856
Light discontinued 1950
Lat / Long: 49°10.90N 2°06.61W
Access inside tower: No

Access to site: The tower can be seen outside the entrance to St Helier Maritime Museum, which is on New North Quay in St Helier. There are a number of public car parks in St Helier (parking disc required) from which it is possible to reach the light on foot.

DEMIE DE PAS

25.20 DEMIE DE PAS (r)

South of St Helier, Jersey

◯ Visited on

....

Black tower, yellow top
Established: 1904
Admiralty number: A1598
Light characteristic: Mo(d)WR 12s
Lat / Long: 49°09.01N 2°06.14W
Access inside tower: No

Access to site: The lighthouse is visible from Poole or St Malo ferry to St Helier. Also visible from a distance from Le Mare car park off of the A4. A closer view would require a chartered boat.

25.21 NOIRMONT POINT (i-t)

West entrance of Saint Aubin's Bay, Jersey

◯ Visited on

....

Black tower, white band (10 metres)
Established: 1915
Admiralty number: A1616
Light characteristic: Fl(4)W 12s
Lat / Long: 49°09.92N 2°10.08W
Access inside tower: No

Access to site: Public parking at the end of Le Chemin de Noirmont. From the car park a path to the south east (right of the old gun station) takes you along a path leading to the lighthouse at low tide. The walk involves a downhill stretch on a rough path and the last section is across rocks where care should be taken, particularly in wet conditions. It is advisable to check tide times before making the journey across to the lighthouse.

25.22 CORBIERE (i-t) ☆

○ Visited on

....

On south west point of Jersey

White round stone tower (19 metres)
Established: 1874
Designed by: Sir John Coode
Admiralty number: A1620
Light characteristic: Iso WR 10s
Lat / Long: 49°10.79N 2°15.01W
Access inside tower: Tours of the lighthouse can be arranged, although at the time of going to print tours were not permitted.

Access to site: Parking for the lighthouse is just off of La Rue de la Corbiere. The lighthouse is visible from the road and the car park. From the car park it is a short walk to the causeway, which is exposed at low tide. The causeway takes you to the lighthouse. Before crossing the causeway, it is best to check tide times to ensure you are able to return safely.

LIGHTHOUSE INDEX